Solar-Powered Urbanization for a Sustainable Future

Solar-Powered Urbanization for a Sustainable Future presents a comprehensive and integrated approach to utilizing solar energy as a key solution for reducing grid dependency and pollution while fostering self-sufficient and cleaner urban centres. This book explores the strategic planning and application of solar power across essential urban sectors, including lighting, heating and cooling, water supply, wastewater recycling, transportation, and urban farming. By analysing energy consumption patterns in residential, commercial, institutional, and industrial spaces, this book highlights sector-specific opportunities for sustainable transformation. With a strong emphasis on awareness, policy formulation, and technological advancements, it contributes valuable insights to the growing discourse on renewable energy integration in urban planning. It serves as an essential resource for researchers, policymakers, and urban planners striving for a solar-powered, sustainable future.

- Explores the role of solar power in key urban sectors to reduce grid dependency and environmental impacts.
- Analyses energy consumption patterns in residential, commercial, institutional, and industrial spaces, identifying tailored solar-powered strategies for sustainable urbanization.
- Highlights the importance of awareness, governance, and policy-making in driving large-scale adoption of solar energy for self-sufficient and environmentally friendly urban centres.

Solar-Powered Urbanization for a Sustainable Future

Arun Kumar and
Sandhya Prajapati

CRC Press
Taylor & Francis Group
Boca Raton London New York

CRC Press is an imprint of the
Taylor & Francis Group, an **informa** business

Designed cover image: Shutterstock

First edition published 2026
by CRC Press
2385 NW Executive Center Drive, Suite 320, Boca Raton FL 33431

and by CRC Press
4 Park Square, Milton Park, Abingdon, Oxon, OX14 4RN

CRC Press is an imprint of Taylor & Francis Group, LLC

ISBN: 978-1-041-09870-6 (hbk)
ISBN: 978-1-041-09892-8 (pbk)
ISBN: 978-1-003-65222-9 (ebk)

DOI: 10.1201/9781003652229

Typeset in Times
by codeMantra

Contents

Preface

Urbanization primarily focused towns or cities that provide industrial and service opportunities for landless labourers; however, it soon incorporated infrastructural development such as improved roadways, designated public spaces, and commercial establishments that significantly transform rural or suburban locales into expansive urban environments. It often facilitates better financial opportunities and new perspectives to lifestyles such as improved public spaces and utilities, large-scale production, and availability and ambition to live luxury life influenced by the capitalistic economic approach which leads to rapid increase in energy consumption.

The increasing energy demands led to the utilization of other energy sources like coal, followed by oil, gas, and later hydropower, nuclear energy, and renewable sources like solar and wind which received attention, thus gaining the momentum to secure energy security. Among the renewable sources, solar energy proved to be the most convenient and affordable energy resource that can fulfil the increasing energy consumption in urban areas. Several approaches are being made to utilize the solar energy developed to tap into this energy source, including solar thermal (heating/cooling), solar power plants, photovoltaics, solar architectural designs, and solar trees.

Solar-powered urbanization is an integrated approach for using solar power for meeting the increasing energy demands of the urban sector that includes residential and commercial, industrial, and agriculture and food processing sectors. However, there are several sub-sectors such as lighting, HVAC (heating, ventilation and air-conditioning), water supply and water recycling, transportation, and agriculture, especially urban farming that are integral part of all the sectors from low to high consumption and also varying or might be specific for a particular sector. These sub-sectors have a rapidly increasing energy consumption profile and will require more energy consumption over the years that need to be addressed to reduce the carbon emissions, mitigate the climate change, and ensure energy security. In order to promote local energy production rather than just consumption in cities, it is essential to incorporate solar energy considerations early in urban design and planning. The implementation of solar solutions in urban areas encounters various challenges that necessitate creative approaches.

The integration of solar systems into urban architecture introduces aesthetic and design difficulties, with public opposition often arising from worries about visual effects. Furthermore, urban power grids might be unprepared to accommodate the distributed energy contributions from solar systems, which call for upgrades and modernization, alongside the need for energy storage solutions to balance supply and demand, particularly during night time. The incorporation of Internet of Things (IoT) devices and data analytics platforms enables intelligent energy usage and grid administration. Smart meters and sensors provide crucial information on energy consumption patterns, allowing city officials to pinpoint opportunities for enhancement and optimization.

About the Authors

Arun Kumar is currently working as an Assistant Professor at Dev Bhoomi Uttarakhand University, Dehradun, in the field of forest restoration, bioremediation, and sustainable environment. He has published various research and review papers in reputed journals such as *Frontiers in Microbiology, Frontiers in Biosciences*, and *Environment and Sustainability Indicators* in the areas of microalgal bioremediation, biofuels, and bioproducts production from microalgae. He is the author of *Microalgae in Waste Water Remediation* and *Solar Powered Wastewater Recycling*, both are published by CRC Press, Taylor & Francis Group. He earned his doctoral degree in Environmental Microbiology from Babasaheb Bhimrao Ambedkar University, Lucknow, India. In his doctoral work, he successfully metabolized a widely used and hazardous insecticide, Chlorpyrifos, by the cyanobacterial strain *Oscillatoria* sp. CYA8 CPF isolated from paddy fields.

Sandhya Prajapati is currently working at DIT University, Dehradun, in the field of solar photovoltaic applications. She has published research and review papers in several leading journals like *Energy Sources, Part A: Recovery, Utilization, and Environmental Effects*. She is the co-author of *Solar Powered Wastewater Recycling*, published by CRC Press, Taylor & Francis Group. She earned her doctoral degree in Electrical Engineering from the Indian Institute of Technology, Roorkee, India. In her doctoral work, she successfully studied the utilization of solar energy in various fields, especially in greywater recycling.

1 Introduction

URBANIZATION AND ITS ASPECTS

Currently the global population is increasing at an annual rate of 1.5%, and improvements in living standards have led to a rise in life expectancy and a decrease in child mortality rates, contributing to an anticipated population surpassing 9.7 billion by 2050 (UNDESA Population Division, 2022). Urbanization is regarded as the development of industrial or commercial hubs within more densely populated regions, characterized by enhanced infrastructure and public services, yet today it represents an ongoing trend of urban modernization and economic transformation marked by consumerism and commercialism (Seto et al., 2022). Initially, this movement was primarily focused on the migration of landless labourers from agriculture-reliant rural regions to towns or cities centred on industrial and service economies, but it soon incorporated elements of structural development, such as improved roadways, designated public spaces, and commercial establishments that significantly transform rural or suburban locales into expansive urban environments (UN-Habitat, 2023) (Figure 1.1).

These new strategies and characteristics of urbanization contribute to unsustainable economic growth, which inevitably causes an increase in energy consumption for both residential and commercial sectors. The heightened consumer behaviour among residents drives industrial sector growth, resulting in greater energy demand to meet these requirements, which consequently leads to exponential rises in carbon

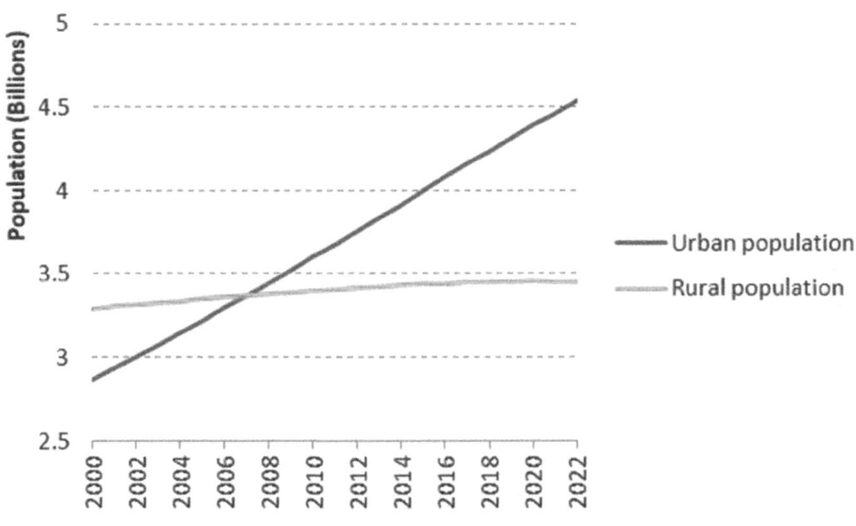

FIGURE 1.1 World population urban vs. rural.

DOI: 10.1201/9781003652229-1

emissions and exacerbates global warming and climate change (Nguyen and Ngo, 2019; Khan et al., 2025). Examining the relationship between urbanization and energy usage provides valuable insights into global population behaviours, such as increased ownership of household appliances, frequent trips to shopping centres (like malls and supermarkets) and higher vehicle ownership. Apart from individual behaviour, society is also actively engaging in an industry-focused economic model, modern agricultural practices, especially in food processing; that are linked to most significant and critical issues such as urbanization assisted increasing energy usage and higher consumption assisted environmental pollution in recent times (Sbia et al., 2017; Shahbaz et al., 2017) (Figure 1.2).

The growing urban energy footprint and resulting environmental degradation have become pressing issues within the world's important economies. Currently most of the working population consider war or conflicts, terrorism, and unemployment as the major threats for humanity in the 21st century, but as compared to affected regions and number of population are in verge of hunger, water scarcity and property damages caused by unpredictable and permanent effects by global warming and climate change (Charfeddine et al., 2018; IPCC, 2023; Khan et al., 2025). This issue is a substantial concern worldwide due to its escalating impacts on human life, exemplified by increased desertification, rising sea levels, and detrimental effects on agriculture, particularly in developing countries. Nonetheless, the intensification of global warming has been linked to multiple factors, including energy consumption, which generates CO_2 emissions—the primary greenhouse gas (GHG) in terms of volume—since the World Resources Institute (WRI) (2024) estimated that CO_2 emissions account for 76.7% of total GHGs. The repercussions of CO_2 emissions are severe, occurring simultaneously with industrial growth and the increasing openness of economies (Martínez-Zarzoso and Maruotti, 2011; National Oceanic and Atmospheric Administration [NOAA], 2025).

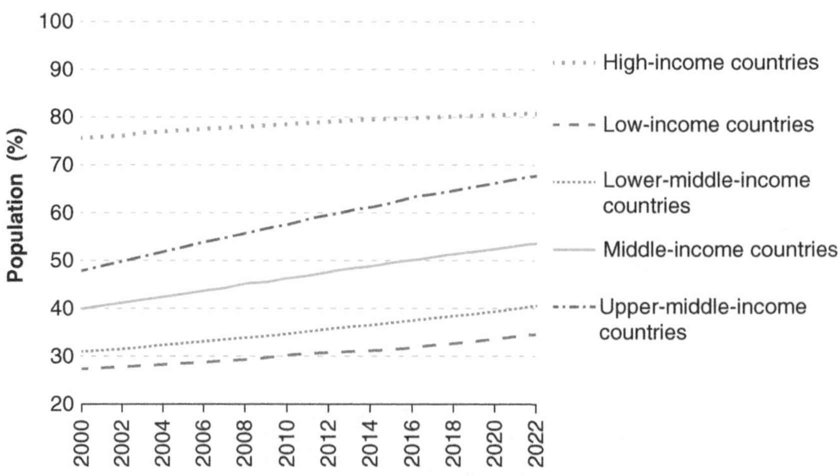

FIGURE 1.2 Share of population living in urban areas.

URBANIZATION AND ENERGY CONSUMPTION

Urbanization often facilitates better financial opportunities and new perspectives to lifestyles such as improved public spaces and utilities, large-scale production, and availability and ambition to live luxury life influenced by the capitalistic economic approach which thrived on conversing (practically altering) the natural resources to value added commodities. The process of urbanization significantly affects energy demand as urban areas expand, leading to rapid increase in residential energy consumption and also elevating energy needs through building construction, transportation facilities, and public infrastructure projects.

This capitalistic and consumerism-based lifestyle rising with urbanization; resulted in the increased per capita energy consumption that can be categorized into three ways:

a. Transition in energy consumption pattern from traditional energy resources to modern fossil fuels to electric vehicles (EVs)
b. Promotion of more demand for goods and services towards technological improvements
c. Elevation of residential and transport related energy consumption

Based on the studies related to urbanization and associated parameters, residential energy utilization, ownership of mobility solutions and industrial and commercial supported production and consumption found that urbanization facilitated the elevation in residential and transport energy consumption. The increased energy consumption further inspires urbanization as more people have ambition to live urbanized life for various personnel, societal, and economic reasons but there are always some rational aspects which inspire the urbanization process.

POPULATION SIZE

Based on various studies, it is established that urban population growth plays a critical role in the rise of energy consumption (Jorgenson and Clark, 2010); as an increasing population needs more residential complexes, mobility solutions including individual, industry-based employment, and intensive agricultural approach all of which require substantial energy. This global trend of population growth in urban areas led to energy consumption correlates significantly with exploitation of natural resources and environmental destruction (Mishalani et al., 2014). Yang (2013) and Liu and Liu (2017) reported that rapid economic expansion and population growth resulted in an increase in urban residential energy use and increased carbon emissions, emphasizing the necessity to control population growth and promote energy efficiency.

ECONOMIC DEVELOPMENT

The technological advancements in energy sources, industrial upgradation, and improved efficiency helps in reducing energy intensity per unit of output (Ma and Ye,

2016). But it also encouraged people to own more appliances which led to increased total residential energy consumption and requiring more products to fulfil the demand led to increased total residential energy consumption. Beside this the level of energy infrastructure and availability of energy supply also responsible for increasing the operating duration of appliances led to increased energy consumption. The consumption behaviour of residents also determines the energy consumption patterns which are primarily associated with comfort and luxury lifestyle (Asif et al., 2024).

INDUSTRIAL STRUCTURE

Meadows and Randers (2012) and Narayanan and Sahu (2014) confirmed that changes in industrial structure have a substantial impact on energy use. In the early stages of industrialization, energy-intensive sectors were prevalent, which heightened overall energy demand. It is also observed that the percentage of high-energy-intensive industrial sectors is directly proportional to energy consumption and carbon emissions. As most of the nations evolved from primarily agriculture dependent economics to limited manufacturing of different value-added products and large-scale industrial production with commercialization of every need, there is a gradual transition towards resource-efficient and knowledge-intensive sectors, fostering intersectoral connections that can help reduce overall energy demand.

TECHNOLOGICAL PROGRESS

The technological advancements and energy consumption are interconnected, it leads to development of cutting-edge technologies that further enhance energy efficiency and lower per capita energy consumption output units (Popp, 2001). Zhou et al. (2017) cautioned that depending solely on technological progress might prove to be contradictory and it leads to elevate the overall energy demand. This complexity suggests that the overall impact of technological advancements on energy consumption can be either beneficial or detrimental. The technological advancements have been considerably helpful in the reduction of energy-related carbon intensity in which energy efficiency increases that might lower the energy costs and improve accessibility, but it also leads to the "rebound effect" resulting in increased total energy consumption (Feng and Ye, 2015).

INDIAN URBANIZATION AND ENERGY CONSUMPTION TRENDS

India ranks as the world's third-largest energy consumer and accounted for around 7.33% of global anthropogenic GHG emissions in 2023 (Crippa et al., 2023). From 2023 to 2050, an anticipated 416 million individuals (representing a 90% increase) will be expected to join the urban population in India (WRI INDIA, 2023); which will lead to increased energy demand and GHG emissions as urban areas grow and smaller towns become more developed. In 2023, India's total GHG emissions stood at 4133.54 Mte CO_2 eq. (third GHG emitter in the world) while per capita GHG emissions 11.57 metric tons Mte CO_2 eq. (the 125th in the world) (Crippa et al., 2023), with almost three times higher in 2023 as compared to 1990. Thus, India's urbanization presents a crucial opportunity to mitigate energy demand and carbon emissions if effective low-carbon city planning strategies are implemented as urbanization progresses.

INCREASING ENERGY CONSUMPTION

India's energy sector is experiencing significant growth, mirroring the rapid increase in the country's population. Over the past decade (from 2013–14 to 2023–24), the total primary energy supply has surged by 54.5%, climbing from 589 Mtoe to 910 Mtoe. Electricity usage, including captive generation, has doubled from 874 Billion Units (BU) to 1,543 BU during this time, reflecting an annual growth rate of 5.8%. The peak electricity demand in India has risen by 79%, growing from 136 GW to 243 GW between 2013–14 and 2023–24. Notably, on May 30, 2024, India reached an all-time high for peak electricity demand at 250 GW.

NEED OF RELIABLE AND SUSTAINABLE ENERGY INFRASTRUCTURE

Over the last ten years, India has reduced electricity shortages by developing a robust clean energy infrastructure. The country's total installed electricity capacity has nearly doubled from 248 GW in 2013–14 to 442 GW in 2023–24 (CEA b, 2024). Additionally, renewable energy capacity has seen more than a twofold increase, growing from 76 GW to 191 GW during the same period. This surge has resulted in a significant rise in the share of renewable energy in India's electricity generation mix, increasing from 18% to 21% in 2023–24. Rooftop Solar (RTS) has experienced impressive annual growth of 45% since 2017–18, expanding from 1.06 GW in 2017–18 to 14.45 GW by 2024–25 (as of October 2024). Peak demand shortages have drastically decreased from 4.5% to 1.4%, while overall electricity requirement shortages have been minimized to 0.3%.

DE-CARBONIZATION IN THE INDUSTRIAL SECTOR

The industrial sector in India is the leading consumer of energy and electricity, accounting for 49.5% and 42% of energy and electricity consumption, respectively, in 2023–24. Over the past decade, there has been a significant shift towards cleaner fuels in the industrial sector, reducing coal's share of total energy consumption from 70% in 2013–14 to 63% in 2023–24. The implementation of energy efficiency initiatives has led to considerable energy savings in the industrial sector, amounting to 25.81 Mtoe and lowering CO_2 emissions by 110.95 MTCO2e annually. The Iron and Steel Sector has effectively achieved its emission intensity reduction targets, cutting emission intensity from 3.1 T/tcs (tonne of CO_2/tonne of crude steel) in 2005 to 2.5 T/tcs in 2020.

TRANSPORT SECTOR FUEL DEMAND AND EVS LED ELECTRICITY CONSUMPTION

The transport sector is the second-largest energy consumer after industry, comprising 12% of final energy usage. This sector plays a vital role in the socio-economic development of the country by connecting 1.4 billion people. The transport sector has seen the greatest increase in energy consumption in India, rising from 35 Mtoe to 72 Mtoe between 2013–14 and 2023–24. This growth is primarily driven by the increased use of vehicles in road transport. Responsible for 87% of passenger traffic and 60% of freight traffic, road transport made up over 77% of the total energy consumption linked to transport in 2023–24. Recently, there has been a noteworthy shift towards EVs, with their share in new vehicle sales growing from 0.3% in 2016–17 to 7% in 2023–24. In the past decade, India has recorded a total of 39.6 lakh EVs, with 11.6 lakhs added between April 2024 and October 2024.

INCREASING RESIDENTIAL ELECTRICITY CONSUMPTION

The residential sector is considered as the second-largest electricity consumer in India which constitutes 32% of total electricity usage. The rise in living standards and large-scale electrification efforts, have led to a significant increase in electricity consumption in residential buildings. From 2013–14 to 2023–24, electricity use rose from 200 BUs to 375 BUs, holding the second-largest share (after industry) in specific electricity consumption in India. The Pradhan Mantri Ujjwala Yojana scheme has greatly boosted the use of Liquid Petroleum Gas (LPG) for cleaner cooking, raising its share of residential consumption from 69% in 2013–14 to 99%.

SOLAR ENERGY UTILIZATION IN AGRICULTURE

Agriculture in India plays a crucial role, supporting nearly 45.76% of the population. The adoption of solar energy for irrigation is fostering a more sustainable and cost-effective method of enhancing agricultural activities in India. The agricultural sector has consistently reduced its oil consumption, with electricity becoming the primary energy source, representing 97% of total energy use in 2023–24. The number of operational pump sets grew from 18.9 million in 2012–13 to 27.2 million in 2022–23, reflecting an annual growth rate of 4%. During the same timeframe, the use of solar pumps expanded significantly, increasing from 11,626 to 0.527 million, marking an extraordinary 45-fold growth.

CURRENT TREND OF RENEWABLE ENERGY WITH FOCUS ON SOLAR ENERGY

Due to population growth, urbanization, and technological innovations, the demand for electricity in India has been rising and is anticipated to grow further in the future. India's power generation sector is experiencing tremendous growth utilizing diverse energy resources including coal, natural gas, hydroelectric, nuclear power, wind energy, and solar energy, along with biomass energy from agricultural and household waste. Coal, oil, and natural gas used for electricity production account for a third of global GHG emissions. Therefore, it is crucial to offer cleaner, more reliable electricity for improving the air pollution conditions; for that there are various energy sources, both conventional and renewable, that are utilized in India's electricity market.

The Indian government aimed to achieve an addition of 175 GW of installed renewable-energy capacity by 2022, with intentions of deriving 100 GW from solar, 60 GW from wind, 10 GW from biomass, and 5 GW from small hydropower plants. Although the target was not achieved, India recorded a significant milestone with 110 GW of installed renewable-energy capacity, which includes 54 GW of solar power. Currently, the total renewable energy capacity amounted to 220 GW (including energy from large hydro power) with 106 GW of solar and 50 GW from Wind energy. Further India revised its renewable energy target 500 GW by 2030 which also includes energy from large hydro power. Although it is an ambitious target like the previous one, analysing recent trends from the year 2022–24; will be promising and achievable with providing the incentives for domestic PV production and involving public–private partnership to ramp up the renewable energy capacity installations in a time bound manner.

The increasing demand for electricity and supportive government initiatives are attracting foreign investments. There has been a rising trend in solar energy production, driven by numerous policies introduced by the Government of India aimed at benefiting the populace and decreasing reliance on electricity consumption. The solar energy generated following these developments will be utilized across various sectors, including industry, information technology parks, and agriculture. Solar water pumps and reservoirs are primarily employed to irrigate agricultural land and store water for irrigation purposes, representing the key applications of solar energy within the agricultural sector.

SOLAR-POWERED URBANIZATION AND ITS COMPONENTS

Urbanization powered by solar energy represents a comprehensive strategy for harnessing solar power to satisfy the growing energy needs of the urban sector, which encompasses residential, commercial, industrial, and agricultural food processing domains. However, there are various sub-sectors involved: (a) Lighting; (b) HVAC (Heating, Ventilation, and Air Conditioning); (c) Water supply and water recycling; (d) Transportation; (e) Agriculture, particularly urban farming; each is crucial across all consumption levels and may also be tailored to specific sectors. These sub-sectors are experiencing a rapid rise in energy consumption, necessitating interventions to curb carbon emissions, combat climate change, and ensure energy security in the coming years.

Solar-powered lighting offers numerous benefits for both the environment and public health. The prevalent use of kerosene lamps, still common in many areas, emits several pollutants such as black carbon, which is closely linked to climate change and contributes approximately 240 million tons of CO_2 emissions annually. The challenges associated with kerosene lamps include potential fire risks and toxicity, primarily due to elevated levels of harmful particulates (Mills, 2016). Concurrently, the growing adoption of renewable energy sources has led to an increased demand for home energy storage systems designed for electricity produced by photovoltaic (PV) systems as costs decline. These systems enable the storage of surplus energy generated by solar or wind sources, making it accessible for household needs at any time. A rising number of private homes are integrating storage systems to temporarily retain electricity produced from PV installations, allowing them to utilize renewable energy for their electronic devices and appliances instead of relying on the grid. Significant advancements are being made in smart lighting technologies. Solar lighting fixtures can now be automated, controlled, and monitored remotely through the incorporation of Internet of Things (IoT) innovations. The implementation of smart sensors, such as light, motion, and temperature detectors, markedly enhances the adaptability of lighting solutions. This advancement not only improves user convenience, but also further enhances energy efficiency.

Traditional heating, ventilation, and air conditioning systems predominantly depend on electricity generated from fossil fuels, which greatly contributes to carbon emissions. In comparison, solar-powered heating and cooling systems harness the endless energy of the sun, transforming sunlight into electricity through PV panels. This renewable energy source reduces our dependence on non-renewable resources and diminishes environmental damage, making it an attractive option for eco-friendly heating and cooling solutions. Solar-powered HVAC systems are made up of solar panels, inverters, batteries, and HVAC equipment. Solar panels, typically installed on rooftops or other exposed locations, capture sunlight and convert it into direct current (DC) electricity. DC converters are frequently employed to transform DC power into the alternating current (AC) needed by the building's electrical system. Any surplus energy can either be stored in batteries for later use or returned to the grid through net metering programs. By integrating solar power with heating, ventilation, and air conditioning systems, energy consumption is significantly reduced. Heat pumps, which are essential components of these systems, utilize solar energy to either heat or cool air, thereby alleviating the demand on conventional heating and cooling systems. Smart thermostats and controls further enhance energy efficiency by adjusting temperatures based on real-time data, thus minimizing energy waste.

Renewable energy-based water management systems are heavily influenced by the use of solar panels. Applying these technologies in facilities for water treatment, irrigation, and pumping, along with energy-efficient solar desalination techniques, promotes sustainable water management practices and addresses challenges related to water scarcity. Progress in solar technology increasingly offers a positive direction for the effective and environmentally friendly oversight of water resources. Access to clean water is becoming a central concern, and solar-powered water treatment plants are emerging as a viable option to satisfy the rising demand. By utilizing solar energy, these facilities minimize their dependence on traditional energy sources, resulting in water treatment systems that are more sustainable for the environment. The integration of solar panels into water and irrigation pumping systems improves water resource management by increasing sustainability and efficiency.. Research led by Vishnupriyan et al. in 2022 investigated the ideal tilt angle and orientation for optimizing solar energy utilization in water pumping systems located in remote areas. Innovations in energy-efficient solar desalination methods have transformed sustainable approaches to water desalination. These groundbreaking solutions tackle the issues of freshwater scarcity and provide more environmentally friendly alternatives than traditional desalination techniques powered by solar energy.

The future of transport is projected to be dominated by EVs due to the rising costs of traditional fuel sources like petrol and diesel. Consequently, the energy demand will surge as EVs are integrated into public transit systems. This review paper examines a solar-powered charging station for EVs by analysing how adjusting the tilt angle of solar panels influences maximum efficiency and power generation from sunlight, factoring in various solar radiation wavelengths. PV solar panels can be employed to charge EVs in an eco-friendly manner. Utilizing solar energy potential in office buildings for charging EVs during work hours, leveraging the extended duration vehicles spend parked, facilitates the implementation of vehicle-to-grid technology. This paper evaluates and compares different architectural designs for an EV-PV charging system. EVs are becoming increasingly accepted in many countries around the globe. Compared to vehicles with internal combustion engines, electric cars are proving to be more energy-efficient and impact the environment less. The limited number of charging stations presents a significant obstacle to the global acceptance of EVs. As the use of EVs grows, more public spaces are beginning to install charging stations. However, if EVs are charged through an existing utility grid powered by fossil fuels, it affects the distribution system and may not be environmentally sustainable. Creating electricity from solar-powered PV panels holds great promise, making it a remarkable option for charging EVs and a sustainable approach to protecting the environment.

Solar energy, a widely available and versatile resource, is gaining traction in a variety of sectors, including residential, industrial, agricultural, and technological fields. This technology can assist in conserving water resources, reducing dependence on the electrical grid, lowering energy costs, and generating additional revenue. Although solar energy has long been utilized in agriculture, there is an increasing demand for its application in a wider array of functions. The agricultural sector faces numerous challenges, such as water shortages, rising energy costs, and the need for sustainable practices; thus, solar energy offers a flexible and dependable solution. Amid the challenges of global climate change and increasing energy demands, the agricultural industry is increasingly turning to renewable energy sources for sustainable alternatives. Solar energy has demonstrated itself as a valuable asset, allowing farmers to reduce expenses, enhance productivity, and promote environment-friendly methods. This comprehensive guide will examine the different ways in which solar energy is revolutionizing agriculture, its advantages, and how it can be utilized to boost productivity in farming. The adoption of solar energy is increasingly being adopted in agriculture.

CURRENT AND FUTURE POLICIES

The International Solar Alliance, established in 2015 by India and France, serves as a prime illustration of how collaborative efforts can guide the global community towards a cleaner and more sustainable tomorrow. This member-driven platform focuses on strategies aimed at creating 450 GW of renewable energy by 2030. It seeks to boost the widespread adoption of solar technologies by broadening energy accessibility, assuring energy security, and facilitating the energy transition among its member nations.

PRODUCTION LINKED INCENTIVE SCHEME

The Government of India's dedication to solar energy is expressed not only through statements but also through detailed policies and programs. The Production Linked Incentive Scheme for the National Programme on High-Efficiency Solar PV Modules was launched to enhance manufacturing capabilities on a Giga Watt scale, with a budget of INR 24,000 crore aimed at promoting domestic production and cementing India's status as a leader in solar energy.

SOLAR PARKS SCHEME

The solar parks offer suitable, developed land equipped with all necessary clearances, transmission infrastructure, access to water, road connectivity, and communication networks. This scheme accelerates the implementation of grid-connected solar power projects on a large scale for electricity production. Each solar park is expected to have a capacity of 500 MW or more, although smaller parks may also be considered in regions where acquiring contiguous land is challenging due to difficult terrain or a severe lack of non-agricultural land. The development of solar parks is carried out in partnership with State Governments, their associated agencies, Central Public Sector Undertakings (CPSUs), and private companies.

PM-KUSUM INITIATIVE

The PM-KUSUM initiative, which aims to achieve an addition of 30.8 GW in solar power capacity by March 2026, is revolutionizing the agricultural sector in India by establishing decentralized solar energy facilities, replacing diesel-powered agricultural pumps with solar-powered ones, and solarizing existing grid-connected pumps. The guidelines of the scheme mandate the use of solar modules, solar cells, motor pump sets, controllers, and the balance of the system to be manufactured domestically.

PM SURYA GHAR: MUFT BIJLI YOJANA (PMSGMBY)

The Rooftop Solar Programme for the residential sector and the Off-grid Solar PV Applications Programme for rural areas are enhancing the accessibility of solar energy through subsidies. This support represents not merely financial aid but a strategic investment towards a future powered by solar energy. The PM Surya Ghar: Muft Bijli Yojana (PMSGMBY), noted as the world's largest domestic rooftop solar program, reached a significant milestone with 1 million homes powered by solar energy as of FY 2025.

REFERENCES

Asif, M., Li, J.Q., Zia, M.A., Hashim, M., Bhatti, U.A., Bhatti, M.A. and Hasnain, A. 2024. Environmental sustainability in BRICS economies: The nexus of technology innovation, economic growth, financial development, and renewable energy consumption. *Sustainability* 16(16), p. 6934.

CEA (2024). National Electricity Plan. Volume II. https://powermin.gov.in/sites/default/files/uploads/National_Electricity_Plan_Volume_II_Transmission.pdf

Crippa, M., Guizzardi, D., Pagani, F., Banja, M., Muntean, M., Schaaf E., Becker, W., Monforti-Ferrario, F., Quadrelli, R., Risquez Martin, A. and Taghavi-Moharamli, P., 2023. GHG emissions of all world countries, Publications Office of the European Union, Luxembourg, 2023, 10, p. 953322.

Charfeddine, L., Al-Malk, A.Y. and Al Korbi, K., 2018. Is it possible to improve environmental quality without reducing economic growth: Evidence from the Qatar economy. *Renewable And Sustainable Energy Reviews* 82, pp. 25–39.

Feng, F. and Ye, A.Z. 2015. Does the rebound effect exacerbate the rise in China's total energy consumption? *The Journal of Quantitative & Technical Economics* 8, pp. 104–119.

IPCC. 2023. Climate Change 2023: Synthesis Report. Contribution of working groups I, II and III to the Sixth Assessment Report of the Intergovernmental Panel on Climate Change. IPCC. https://www.ipcc.ch/report/ar6/syr/

Jorgenson, A.K. and Clark, B., 2010. Assessing the temporal stability of the population/environment relationship in comparative perspective: A cross-national panel study of carbon dioxide emissions, 1960–2005. *Population and Environment* 32(1), pp. 27–41.

Khan, U., Khan, A.M., Alam, M.S. and Houaneb, A. 2025. Toward sustainability: Understanding the impact of economic growth, urbanization, energy use, and resource management on carbon emissions. *International Journal of Energy Economics and Policy* 15(2), pp. 637–646. https://doi.org/10.32479/ijeep.18846

Liu, M.Z. and Liu, X.X., 2017. Study on the influencing factors of urban household energy consumption in China based on STIRPAT Model. *Resources and Environment in the Yangtze Basin* 26(08), pp. 1111–1122.

Ma, L. and Ye, Q.Q. 2016. Empirical research on the relationship between energy consumption and economic growth—A case study of Shaanxi province. *Economic Geography* 36, pp. 130–135.

Martínez-Zarzoso, I. and Maruotti, A. 2011. The impact of urbanization on CO2 emissions: Evidence from developing countries. *Ecological Economics* 70(7), pp. 1344–1353.

Meadows, D. and Randers, J. 2012. *The limits to growth: The 30-year update*. Routledge.

Mills, E. 2016. Identifying and reducing the health and safety impacts of fuel-based lighting. *Energy for Sustainable Development* 30, pp. 39–50.

Mishalani, R.G., Goel, P.K., Westra, A.M. and Landgraf, A.J. 2014. Modeling the relationships among urban passenger travel carbon dioxide emissions, transportation demand and supply, population density, and proxy policy variables. *Transportation Research Part D: Transport and Environment* 33, pp. 146–154.

Narayanan, K. and Sahu, S.K. 2014. Energy consumption response to climate change: Policy options for India. *IIM Kozhikode Society & Management Review* 3(2), pp. 123–133.

National Oceanic and Atmospheric Administration (NOAA). 2025. Current GHG levels. https://www.climate.gov/ghg/current-levels#:~:text=Methane%20is%20the%20second%20largest,than%20it%20was%20in%201990

Nguyen, T.T.H. and Ngo, L.V. 2019. The influence of urbanization on energy consumption: A study in emerging markets. *Energy Economics* 83, pp. 336–345. https://doi.org/10.1016/j.eneco.2019.07.015

Popp, D.C. 2001. The effect of new technology on energy consumption. *Resource and Energy Economics* 23, pp. 215–239.

Sbia, R., Shahbaz, M. and Hamdi, H. 2017. A contribution of foreign direct investment, clean energy, trade openness, carbon emissions and economic growth to energy demand in UAE. *Journal of Cleaner Production* 161, pp. 849–857. https://doi.org/10.1016/j.jclepro.2017.05.150

Seto, K.C., Sánchez-Rodríguez, R. and Fragkias, M. 2022. Urbanization, global environmental change, and sustainability. *Current Opinion in Environmental Sustainability* 60, p. 101201.

Shahbaz, M., Loganathan, N., Muzaffar, A.T., Ahmed, K. and Jabran, M.A. 2017. How urbanization affects CO_2 emissions in Malaysia? The application of STIRPAT model. *Renewable and Sustainable Energy Reviews* 57, pp. 83–93. https://doi.org/10.1016/j.rser.2015.12.040

UNDESA Population Division. 2022. World Population Prospects 2022: Summary of results. UN DESA/POP/2022/TR/NO.3.

UN- Habitat. 2023. World Cities Report 2022: Envisaging the future of cities. United Nations Human Settlements Programme. https://unhabitat.org/world-cities-report-2022

Vishnupriyan, J., Partheeban, P., Dhanasekaran, A. and Shiva, M., 2022. Analysis of tilt angle variation in solar photovoltaic water pumping system. Materials Today: Proceedings, 58, pp. 416–421.

World Resources Institute. 2024. Initiatives. https://www.wri.org/initiatives

WRI India. 2023. Measuring urbanization: Why India needs to re-think its methodology. https://wri-india.org/perspectives/measuring-urbanization-why-india-needs-re-think-its-methodology

Yang, J. 2013. Research on the relationship between energy consumption, economic growth and carbon emissions based on econometric models. Master's Thesis, Northwest Normal University, Lanzhou, China.

Zhou, J.G., Zhang, J.G., Yan, J.Y., and Du, Q.J. 2017. Is Technological Progress a Panacea for Energy-Environmental Constraints? Observations on Jevons Paradox. Jiangxi Hebei GEO University. 2017, 40, pp. 31–43.

2 Urbanization and energy consumption

URBANIZATION: DEFINITION AND CONCEPT

Urbanization is not just limited to migration of agriculture-dependent rural and suburban workers to industry and service sector for financial security but it could be considered as a tool for up gradation or structural changes in the society and economy. Around the world, the concept of an ‘urban area’ doesn’t have a universal definition, it has region- or nation-wise variations that are primarily attributed to minimum number of persons for an inhabitation or other measures such as population density, infrastructure, and pre-approved urban centres. The minimum number of persons for an inhabitation to be designated as urban settlement further has a vast variation across countries like Sweden and Denmark considering this minimum number of inhabitants at only 200 inhabitants; while Japan at 50,000 inhabitants shows a 250 times difference in inhabitants (Ritchie et al., 2024).

Due to this, there is disagreement over current UN estimates of urbanization, further contended that inconsistent concept and metrics for an “urban” area across regions or nations cause miscalculation either underestimation or overestimation, in deciding the world’s urban population. European Commission, Food & Agriculture Organization (FAO), UN-Habitat, International Labour Organization (ILO), Organization for Economic Co-operation and Development (OECD) and World Bank, collaborated for discussion variations or comparisons and developed a new people-oriented universal definition of concept for urbanization (EUROSTAT, 2021). A new urban concept or definition known as “Degree of Urbanization” to describe the cities, towns, and villages (EUROSTAT, 2021; UN-Habitat, 2022; UN-Statistical Commission [UNSC], 2020, 2022). Further the national census data integrated with high-resolution satellite imagery under the project, Global Human Settlement Layer (GHSL) to depict its estimates of cities, towns, and villages. Based on minimum inhabitants and population density, European Commission presented a systematic definition or criteria of habitation across all their nations (European Commission, Joint Research Centre [EC-JRC], 2018; UN-Habitat, 2022; Ritchie et al., 2024):

For cities, there must have a minimum of 50,000 population with density of at least 1,500 people per square kilometre or density of build-up area greater than 50%.

For towns and semi dense areas, there must have a minimum of 5,000 population with density of at least 300 people per square kilometre.

- Towns: settlements with between 5,000 and 50,000 that are either dense (with a density of at least 1,500 inhabitants per square kilometre) or semi-dense (a density at least 300 inhabitants per square kilometre).
- Suburban or peri-urban areas: cells belonging to urban clusters but not part of a town.

DOI: 10.1201/9781003652229-2

For rural areas including villages, there are fewer than 5,000 populations.

- Villages: settlements with a population between 500 and 5,000 inhabitants and a density of at least 300 inhabitants per square kilometre.
- Dispersed rural areas: rural grid cells with a density between 50 and 300 inhabitants per square kilometre.
- Very dispersed rural areas or mostly uninhabited areas: rural grid cells with a density between 0 and 50 inhabitants per square kilometre.

Based on this criteria and integration of satellite imagery, there were around 45% of the world population residing in cities, 35% residing in towns; while the remaining 20% reside in villages. Following this scenario, the urban population reached up to 80% of the total world population in 2020 (more than 6.2 billion people); while the UN estimation stated that only one-third or less of the population lies in the urban areas (Pesaresi et al., 2024). Although there were contradictions in definition, a city or an urban area led to inaccuracies in its estimated population. However, this contradiction found to be minimal in case of cities with populations of at least 300,000, the estimation found to be closer as 31% of the population residing in cities according to the UN, to 29% of the population residing in cities (EC-JRC, 2018) (Figure 2.1).

Besides having many distinctive differences in the estimation of the numbers at the world scenario, the all-round features of urbanization at national levels (irrespective of their description) is relevant especially for India, where the population in urban areas has recorded tremendous growth and reached almost double of the urban population in the year 1990 (Bhagat, 2018; Dijkstra et al., 2021). This increasing trend of living in urban areas could be helpful to find out the assessment of urbanization progress, changes in demographics, and urban planning and policies at national level. However there is no consensus on the global scenario regarding urbanization and its aspects, especially the numbers, but it is useful from a national perspective to understand the reality of urbanization and the numbers of people who live in urban centres either town or cities (Figure 2.2).

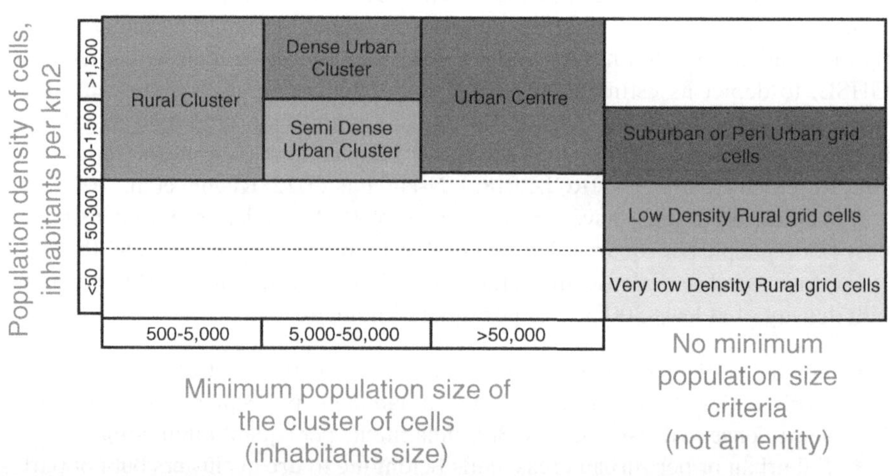

FIGURE 2.1 UN degree of Urbanization.

FIGURE 2.2 (a, b, c) Global population of cities, towns and villages.

POPULATION DENSITY AND LARGE URBAN AGGLOMERATIONS

The population density is undoubtedly a characteristic feature of the urban areas, it also does not differentiate into small and large urban areas; despite having low population density settlements could be considered as cities while some urban areas with high population density could be considered as towns. Further there are significant differences in large urban agglomerations across the world like most of the American nations have 40%–50% chances of having megacities; while most nations across Africa, Asia and Europe showed 10%–40% presence of megacities (UN DESA, Population Division, 2018). UN DESA, Population Division (2018) represented that some nations that considered for presence of smaller cities, such as Japan, Israel, Kuwait, UAE, and Puerto Rico, have more presence of megacities: it is also observed that their maximum populations nearly 50% of the total residing in the large cities; while few nations such as Germany and Poland despite having high rate of urbanization, tend to have low chances of having mega cities, less than 10% of the population residing in large cities.

Over the past five decades, rapid urbanization is a global phenomenon, characterized by significant population growth in numerous cities worldwide. Figure 2.3 displayed the projected population of the world's 30 biggest cities based on the growth of population from 1950 to 2015, ranked by the 2015 population, with forecasts extending to 2,035.9; notably, Beijing in 1950 had a population of approximately 1.7 million. This had surpassed ten times the previous amount by 2015, reaching over 18 million. By 2035, this number is projected to rise to over 25 million. The capital of Bangladesh, Dhaka, expanded from approximately half a million residents in 1950 to almost 18 million inhabitants in 2015, with predictions suggesting a population surpassing 30 million by 2035 (UN DESA, Population Division, 2018; Ritchie et al., 2024).

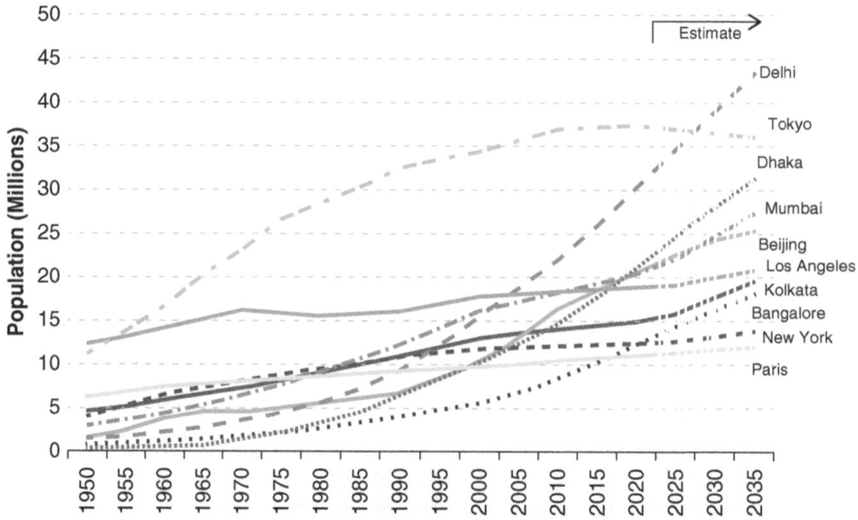

FIGURE 2.3 Population of the world's largest cities 2000–2035.

ISSUE OF SLUMS IN URBAN CENTRES

With increasing population in urban agglomerations, the percentage of people residing in slums rapidly increased, it is estimated that one-in-four urban dwellers residing in slums or slum like areas where group of individuals residing in small space deprived, at least one of the following requirements: access to clean water, access to proper sanitation facilities, sufficient living space, and a durable house. Living standards in urban areas are undoubtedly a significant indicator of overall wellbeing. One indicator of living standards is the proportion of the urban population residing in slum households. The nations across Asia and Latin America had 10–50% of urban populations residing in slums; while most of the nations in Sub-Saharan Africa showed that more than half of the urban populations residing in slums and in some nations such as Chad around eight in ten people residing in slums. Although the share of the urban slum population has been declining over time in most of the countries, it is significant in the number of people forced to live in slum-like conditions.

URBANIZATION AND ENERGY CONSUMPTION TRENDS IN INDIA

The "urban area" India seems to be quite constrained as it includes all the three criteria – i.e. population size, population density and economic perspective; while most of the nations across the world considers just one or two criteria, that might be from administrative, demographic, population density, and economic perspective to physical attributes (population size and built-up) and levels of infrastructure (buildings, roads, and infrastructure). NSO-India (2019) classified settlements in to 'Statutory Towns' which includes urban administrative units such as municipalities, municipal corporations, cantonment boards, or notified town area committees; while Census Towns includes all other inhabitants that fulfil all three criteria—minimum population of 5,000, population density of at least 400 individuals per square kilometre and 75% male individuals engaged in non-agricultural work population.

Considering the current concept and definition, India has only 31.2% of population residing in urban areas in 2010 that was considered far less than the scale of urbanization (Census, 2011); while according to the World Bank's Agglomeration Index, it has predicted up to 55.3% of the population residing in urban-like features in 2010 (World Bank, 2015). Although there are inaccuracies or differences in concepts or criteria in deciding the urban population, if the urbanization trend goes with Indian concepts, there will be an increase from 31% in 2011 to 40% of the population residing in urban areas in 2030.

With increasing urbanization trends, India has also recorded a significant rise in the electricity consumption including all the sectors i.e. residential, commercial, and industrial over the years; resulting in the nearly doubling over the past decade. In the year of 2023–24, there was a 7% increase in the total electricity consumption in India, reaching 1,543 TWh, while per capita electricity consumption also increased to 1,395 kWh (but still very low of the global average, nearly one-third) as compared to previous years (CEA, 2024). The increasing electricity consumption is fuelled by all the sectors, in which the industrial sector, responsible for 42% of the total electricity consumption, is necessary for powering the manufacturing and production activities of the country. The residential sector comprises 24% of total electricity consumption, with a 6% annual increase rate; driven by the improved access to electricity in most of the urban areas and increased use of various electrical appliances especially for cooling purposes. The agriculture sector is responsible for 17% of the total electricity consumption, driven by improved access to electricity in rural areas, and an increased dependency on electric pumps for irrigation of fields (Prayas, 2016; BEE, 2024) (Figure 2.4).

In India, there is an 8% average annual rate in the residential energy consumption while comparing this to other developing countries it is low as China's residential energy consumption increased at a 12% average annual rate. But in the case of more developed countries like South Korea and USA, it just showed a moderate rate of increase (5% and 1% respectively), and Germany showed a constant level of residential energy consumption (Prayas, 2016). By comparing total residential energy consumption not so significant to deciding the trend and behaviour of consumers, there is a need for per capita residential energy consumption. India and China show much lower per capita residential energy consumption as compared to all the developed

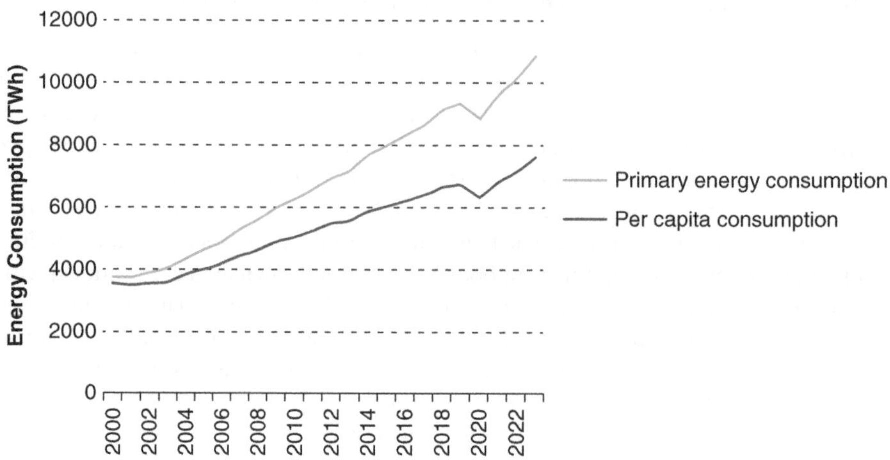

FIGURE 2.4 Total energy consumption with per capita consumption.

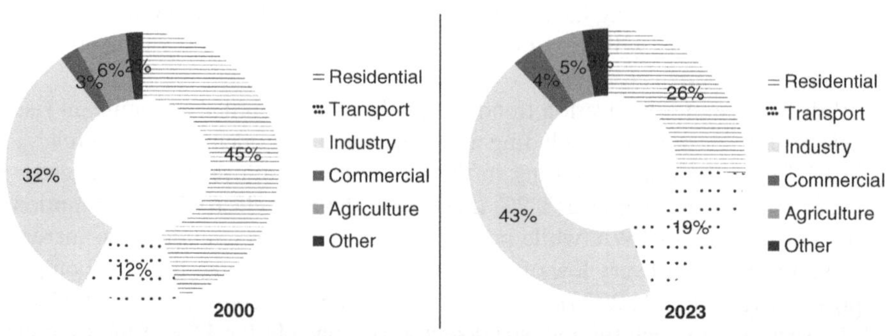

FIGURE 2.5 Sector wise electricity consumption of India (2000 vs 2023).

countries; the USA has about 30 times more per capita residential energy consumption than India, while China has four times more capita residential energy consumption than India (Figure 2.5).

RESIDENTIAL ELECTRICITY CONSUMPTION (REC)

Residential electricity consumption (REC) is the sum of the total electricity consumption by households for lighting, cooking, air conditioning, refrigeration and water heating especially in the winters. Beside that, various appliances like TV, washing machines, water pumps and vacuum cleaners are extensively used; while inverters (Uninterrupted Power Supply [UPS]) are also the integral part of Indian households to store the grid-connected electricity as backup. It is estimated that total REC in India recorded 50 times more than the level in 1971, where the share of REC just accounted for 4% of the total electricity consumption; it rose to 22% of total electricity consumption of 883 billion units in 2014 (CEA, 2011, 2014; MOSPI, 2015).

The factors responsible for increasing REC depend upon the decisions of the households to use more electrical appliances that thrive on a number of factors such as income, tariff, and electricity supply, climate and building design. The residents who have higher incomes tend to purchase and utilize more appliances. For example, almost all the income groups own a TV and fan but air-conditioners are still limited only to residents with higher incomes.

India has different climatic zones (that are primarily categorized on the average temperature and humidity level), that significantly affect the purchase and utilization of heating and cooling appliances; like residents utilize more fans and air-conditioners in hot climates while more water heaters and other space heating appliances in colder climates. Further utilization of these appliances is often quantified by cooling and heating degree days (CDD & HDD) (a measurement of high or low temperature (in degrees) above or below a normal level, and the time period (in days) of these conditions occurs). For a country like India (with tropical conditions), CDDs are prevalent and responsible for more electricity consumption as compared to HDDs which are more resistible (Borah et al., 2015).

Further design of buildings and the choices of construction materials also influence the heating/cooling and lighting needs. As the sunlight glares on a building, or on a particular side of the building, leads to gradual heating up the surface and interior of the building. This situation is primarily managed by installing windows, doors and other such openings to provide cross ventilations that can increase or decrease the heat trapped inside the building. If this is not properly achieved it might lead to the need of air conditioning that escalates the energy consumption.

TRANSPORT ENERGY CONSUMPTION

Currently India has around 1.4 billion people, and progressing on a 7% average economic growth rate over the last 20 years (Chakraborty, 2021), requires the need a robust mobility system to transport the people and to the goods to run economy (NITI Aayog, and RMI India, 2021). According to IEA (2021), the transport sector alone is responsible for 50% of oil consumption in India, where it is used for motorized passenger travel mainly road, rail air travel (small share) 1; while few coastal states used waterways but it is the least-used mode (almost negligible) (Nag, 2021). In India, the people prefer to own and use two-wheelers and four-wheelers for their mobility requirements. As for the people's income increasing in India, the numbers of vehicles, especially ownership of two-wheelers, are showing significant growth (Anup and Yang, 2020). Chaturvedi (2021) estimated that total motorized passenger activity of India surged up to 3,833 billion passenger-kilometres in 2020 from roughly 1,700 billion passenger-kilometre (pkm or passenger-kilometre is a measure of movement of passengers by a mode of transport, like roads, railways, airways or waterways) in 2005.

Beside passenger travel, there is significant growth in freight movement mainly through roads, there is a twofold increase in 2005 levels of passenger travel surged up to 2,250 billion tonne kilometres in 2020. The main reason for the acceleration in freight demand is assisted with the insignificant expansion of the rail–road share (National Transport Development Policy Committee, 2014). In freight movement, trucks or trailers lead a major portion of freight movement, accounting for approximately 65% of the total freight movement in 2020. Further improved road infrastructure through construction of more national highways, has accelerated the speed and the average distance across which freights carried by the trucks.

The transport sector in India accounts for a major portion of energy consumption, especially oil, it is estimated that the transport sector alone is responsible for around 19% of the total energy consumption (around 4 EJ of energy) four and also consumed about 50% of the total oil consumption of India (IEA 2020). By comparing the modes of transport and vehicles, 31% of the total passenger energy consumed by the two-wheelers, followed by 29% by the buses and three-wheelers and 27% by the four-wheelers despite having low fuel efficiency per passenger compared to other modes of transport. Apart from this domestic aviation only contributes for just having 2% share in passenger activity, leading to 8% oil consumption total energy consumption.

With rising income levels of people and on-demand mobility services (such as Ola and Uber), will drive four-wheelers' demand in the next three decades, their share will increase from 9% in 2020 to a staggering 45% in 2050. Further there will be an expectation that electric four wheelers will achieve price equivalence with ICE based four wheelers, will lead to the share of electric four wheelers from 2% in 2020 to 11% in 2050, resulted in the reduction of shares of other modes like two-wheelers and public transport. It is also expected that by 2030, about a third of the four-wheelers and half of the two-wheelers sold might be electric, and this proportion could potentially rise to 75% and 90% by 2050.

Increasing economic development will need significant growth in especially freight transport activities that need to expand. There is an expectation that demand for freight services is expected to increase by a factor of five, from approximately 2,000 billion tonne-kilometres in 2020, to over 10,000 billion tonne-kilometres by 2050. Currently the transport sector largely depends on oil consumption but this will change in the coming decades where the sector's energy mix will also diversify.

Overall, the energy demand of the transport sector will experience more than three times growth from nearly 4 EJ in 2020 to more than 13 EJ in 2050, while the oil consumption in total transport energy needs is projected to decrease from 95%

in 2020 to 66% by 2050. The shift in the fuel mix will arise from an increase in the proportion of natural gas (from 3% to 25%) and electricity (from 2% to 9%).

INDUSTRIAL ENERGY CONSUMPTION

India's industrial energy consumption has the biggest energy footprints, accounting for nearly 49.5% energy consumption in 2023–24. In India, industries are divided into two categories: energy-intensive and non-energy-intensive sectors. Sectors requiring substantial amounts of energy include industries such as iron and steel, petrochemicals, aluminium, cement, pulp and paper, textiles, chlor-alkali, and fertilizer, whereas sectors that use relatively little energy comprise automotive, food processing, pharmaceuticals, and IT, as well as a broad spectrum of micro, small, and medium-sized enterprises. The industry sector relies mainly on coal, oil, and natural gas for its energy needs. The energy production in the industry sector saw coal usage experience a 3% annual growth rate, increasing from 302 MT in the 2016–17 period to 366 MT in the 2023–24 period; oil consumption rose slightly from 31 Mtoe in the 2013–14 period to 52 Mtoe in the 2023–24 period.

Coal's proportion dropped from 70% in 2013–14 to 63% in 2023–24 in total energy usage, whereas oil consumption rose from 14% to 17%. Among oil products, naphtha is the preferred choice, followed by Furnace Oil and then diesel. Naphtha serves as a crucial raw material for the petrochemical industry, comprising 75% of its total consumption in that sector for 2023–24, and is also commonly used in the chemical industry as a solvent. Meanwhile, natural gas consumption experienced a 28% increase from 2018–19 to 2023–24, largely due to rising demand in the fertilizer sector, which accounted for 31% of total natural gas consumption and primarily enabled urea manufacturing in 2022–23.

India's industrial sector accounts for 42% of the total electricity consumption as of 2023–24, making it the country's largest electricity consumer. Industrial sector electricity consumption rose from 440 TWh in 2016–17 to 645 TWh in 2023–24. The rise in this increase can be attributed to the expansion of production capacity, the electrification of other industrial processes, and technological advancements in the sector, although there is a shift towards adopting clean energy over the last decade. The implementation led to a decrease in coal's share from 70% in the 2013–14 periods to 63% in 2023–24 in overall energy consumption; this resulted in substantial energy savings of 25.81 Mtoe and a reduction of CO2 emissions of 110.95 Mt CO2e per year.

AGRICULTURE ENERGY CONSUMPTION

In terms of contributions to India's Gross Value Added (GVA) and job opportunities for the masses, the agriculture sector is seen as extremely crucial for the economy and the people of the country. The agriculture sector accounted for 18.1% of India's GVA at current prices in the fiscal year 2022–23, and is often considered the backbone of the nation's economy. According to the 2011 census, approximately 54.6% of the population in rural areas derive employment from the agricultural sector, although this sector's contribution to India's economy has been declining over the past decade.

Energy usage within the agricultural sector encompasses a range of activities including soil preparation and sowing, crop cultivation, irrigation, harvesting, processing, transportation, and storage of agricultural commodities. The agriculture sector's use of chemical fertilisers, which require a significant amount of energy to produce, also contributes to substantial energy consumption. The agriculture sector requires fuel for tractors, harvesters, or other farm machinery, as well as electricity for irrigation. However, current trends indicate that electricity's share is growing, accounting for 97% of total fuel consumption in 2023–24.

According to de-dieselization efforts and other energy efficiency measures, there has been a 28% decrease in the number of diesel pump sets used from the fiscal years 2016–17 to 2020–21. Improving irrigation access has been a continuous effort over the past few years, with 27.16 millions of pump sets energized by the end of 2022–23. Following a gradual increase in energized pump set installations from 2016 to 2017, a 23% spike in energized pump sets was observed from 2020–21 to 2022–23. The number of solar pump installations has grown at a compound annual growth rate (CAGR) of 28% between 2016–17 and 2022–23. Meanwhile, the agriculture sector's electricity consumption has risen at an average annual rate of 4.2% from 2016–17 to 2023–24. The steady increase suggests a growing requirement for electricity within the sector during this period.

The Government's push to de-dieselize agriculture by 2024 is leading to notable changes in the use of petroleum and fuel-based products. The change is driven by the effort to solarize water pumps and explore alternative non-fossil-based fuels for powering tractors and harvesters. Diesel, encompassing Light Diesel Oil and High-Speed Diesel, is the most commonly consumed fuel, primarily used for powering tractors, harvesters, and threshers. Its consumption declined by 32% from 2019–20 to 2023–24, following a drop-off in usage post 2019–20, attributed to increased electrification of irrigation pumps and above-average rainfall. The second highest fuel source consumed was Furnace oil. Usage increased from 2017–18 to 2020–21, but has since declined to 0.02 million tonnes in 2023–24.

REFERENCES

Anup, Sunitha, and Zifei Yang. 2020. New two-wheeler vehicle fleet in India for fiscal year 2017–18. Working Paper 2020-08, International Council on Clean Transportation.

BEE. 2024. India energy scenario 2023–24. https://beeindia.gov.in/sites/default/files/BEE_India_Energy_Scenario_Report-2024.pdf

Bhagat, R.B. 2018. *Urbanization in India: Trend, pattern and policy issues*. International Institute for Population Sciences.

Borah, P., Singh, M.K. and Mahapatra, S. 2015. Estimation of degree-days for different climatic zones of North-East India. *Sustainable Cities and Society* 14, pp. 70–81.

CEA. 2011. *18th electric power survey of India* (Vol. 1). Central Electricity Authority.

CEA. 2014. *General review 2014*. Central Electricity Authority (CEA).

CEA. 2024. CEA Annual Report 2023–24. https://cea.nic.in/wp-content/uploads/annual_reports/2024/Final_Approved_Annual_Report_2023_24_05032025_1.pdf

Census. 2011 Office of the Registrar General & Census Commissioner, India Government of India. https://censusindia.gov.in

Chakraborty, A. (2021). Present Bias. *Econometrica* 89(4), pp. 1921–1961. https://doi.org/10.3982/ECTA16467.

Chaturvedi, Vaibhav. 2021. Peaking and Net-Zero for India's Energy Sector CO2 Emissions: An Analytical Exposition. New Delhi: Council on Energy, Environment and Water.

Dijkstra, L., Florczyk, A., Freire, S.M.C., Kemper, T., Melchiorri, M., Pesaresi, M. and Schiavina, M. 2021. Applying the degree of urbanisation to the globe: A new harmonised definition reveals a different picture of global urbanisation. *Journal of Urban Economics* 125, p. 103312.

European Commission, Joint Research Centre (EC-JRC). 2018. *Atlas of the human planet 2018, a world of cities*. Publications Office of the European Union.

EUROSTAT. 2021. *Applying the degree of urbanisation a new international manual for defining cities, towns and rural areas 2021 edition.* European Commission. Statistical Office of the European Union.

IEA. 2020. World Energy Balance 2020. Paris: International Energy Agency. https://www.iea.org/subscribe-todata-services/world-energy-balances-and-statistics.

IEA. 2021. Methane and climate change. https://www.iea.org/reports/methanetracker-2021/methane-and-climate-change.

MOSPI. 2015. *Energy statistics 2015* (Vol. 22). Central Statistics Office, MOSPI.

Nag, Devanjana. 2021. Eco-friendly transport! Govt plans more ro-pax services across India's vast coastline & inland waterways. Financial Express, June 08. https://www.financialexpress.com/infrastructure/ eco-friendly-transport-govt-plans-more-ro-paxservices-across-indias-vast-coastline-inlandwaterways/2267507/.

National Transport Development Policy Committee. 2014. India Transport Report: Moving India to 2032. New Delhi: Routledge.

NSO-India. 2019. Seventh Economic Census, 2019. https://www.mospi.gov.in/sites/default/files/economiccensus/seventh_economic_census/Concepts%20&%20Definitions.pdf

Pesaresi, M., Schiavina, M., Politis, P., Freire, S., Krasnodębska, K., Uhl, J.H., Carioli, A., Corbane, C., Dijkstra, L., Florio, P. and Friedrich, H.K. 2024. Advances on the global human settlement layer by joint assessment of Earth observation and population survey data. *International Journal of Digital Earth* 17(1), p. 2390454.

Prayas. 2016. Residential electricity consumption in India: What do we know? https://energy.prayaspune.org/images/pdf/residential_electricity_consumption_in_india.pdf

Ritchie, H., Samborska, V. and Roser, M. 2024. Urbanization. Our world in data. https://ourworldindata.org/urbanization

UN DESA, Population Division. 2018. *World urbanization prospects: The 2018 revision*. Population Division.

UN-Habitat.2022. *World cities report 2022: Envisaging the future of cities*. United Nations. https://unhabitat.org/sites/default/files/2022/06/wcr_2022.pdf

UNSD. 2022. A recommendation on the method to delineate cities, urban and rural areas for international statistical comparisons. https://unstats.un.org/unsd/statcom/51st-session/documents/BG-Item3j-Recommendation.pdf

UN-Statistical Commission. 2020. *Report on the fifty-first session (3–6 March 2020)*. Supplement No. 4 E/2020/24-E/CN.3/2020/37. Economic and Social Council, Official Records.

World Bank. 2015. https://www.worldbank.org/en/country/india/brief/leveraging-urbanization-india

3 Energy consumption and issues

EVOLUTION OF ENERGY PRODUCTION AND GLOBAL ENERGY MIX

Before the middle of the 19th century, the conventional biomass (utilization of solid fuels such as wood, agricultural waste, or charcoal was the dominant energy source worldwide; it continued till Industrial Revolution took place in the 20th century, leading to an increase in the energy demands (Ritchie, 2021). The increasing energy demands led to utilization of other energy sources like coal; that followed by oil, gas, and later hydropower evolved to fulfil the energy needs. Further use of nuclear energy got into the picture during the 1960s. Further there is a need for other energy resources but it could be renewable like solar and wind got attention in the 1980s. Having so much diversification in energy resources, still global energy needs is fulfilled through energy from fossil fuels, i.e. oil, coal, gas, and hydroelectric power (Smil, 2017).

Over the years, the primary energy consumption at world level has surged up rapidly and is expected to reach 646 EJ in the year of 2045; further hydropower and renewable energy consumption are gaining the momentum to reach 30% of the global energy consumption between 2045 and 2050 (Statista, 2025). There are some countries that responsible for most of the primary energy consumption i.e. China, USA, and India are the largest consumers of primary energy globally, while Qatar, Singapore, the United Arab Emirates, and Iceland are responsible for the highest per capita energy consumption. In the last two decades there was a significant progress in renewable energy generation and consumption that has been expanded from 51 EJ in 2000 to reach over 90 EJ in 2025 (Statista, 2025) (Figure 3.1).

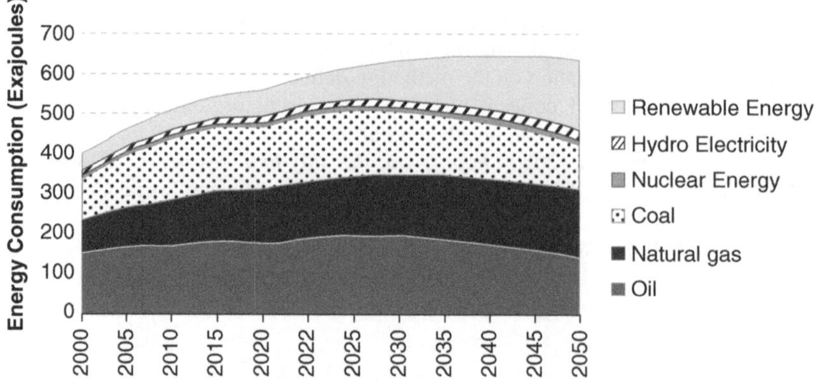

FIGURE 3.1 Global energy consumption prediction up to 2050.

DOI: 10.1201/9781003652229-3

INDIAN ENERGY MIX AND TRANSITION TO RENEWABLE ENERGY

India's increasing population, expansion of energy needs and remarkable economic growth, driven to an increase in the primary energy supply that rose from 589 Mtoe in 2013–14 to 910 Mtoe in 2023–24; that is an increase of 54.5% in the duration of just ten years (BEE, 2024). In the same duration of 2013–14 and 2023–24, the total electricity consumption (including captive power generation) recorded almost double from 874 Billion Units to 1,543 Billion Units with an annual increase rate of 5.8%; while there is significant growth of 79% in India's peak electricity demand that rose from 136 GW to 243 GW during 2013–14 to 2023–24.

It is evident that India's 80% or more energy needs are still supplemented by fossil fuels i.e. coal, oil, and natural gas including solid biomass. Although coal proved to be the largest source for power generation and energy needs for the industries; it will also remain the primary energy in the energy mix over the next few years, the increasing trend of owning more vehicles and road transport use lead to increase in oil consumption and imports. Despite the extensive LPG coverage, the use of solid biomass, primarily fuel wood, is extensively popular as a cooking fuel especially in rural and urban slums, but following a decline trend.

Despite increasing coal and oil demand in India, there is a declining trend of conventional energy resources to achieve a more diversified energy mix through increasing use of renewable sources, leading to a robust clean power infrastructure over the last ten years. During 2013–14 to 2023–24, India's installed electricity capacity almost reached to 442 GW from 248 GW (CEA, 2024); while renewable energy capacity also increased from 76 GW to 191 GW. In India's power generation mix; the renewable energy share surged from 18% to 21% in just ten years; in which renewable energy sector showed a remarkable growth of 45% annually in roof-top solar from the year 2017–18; it surged up to 14.45 GW in 2024–25 from just 1.06 GW. Beside more electricity generation and use of renewable resources, India also shelved the peak demand shortage to 1.4% from 4.5% and the difference between overall electricity requirement and shortage bringing down to just 0.3% (Figure 3.2).

ELECTRICITY LOSSES IN DISTRIBUTION AND TRANSMISSION LINES

There is a significant mismatch between the units of electricity generated at power stations with the units of electricity transmitted and distributed to the consumers, leading to loss of some percentage of the electricity units known as transmission and distribution (T&D) loss. This T&D loss is the unit of electricity compensated by the power generation companies and DISCOMs (the entities responsible for the distribution of the electricity to the consumers) that are not paid for by users. Most of the loss of electricity occurred during the distribution, i.e. approximately 50%, while transmission losses contributed up to approximately 17%. The T&D losses can be classified into two categories: (a) Technical Losses; and (b) Non-Technical Losses (Commercial Losses).

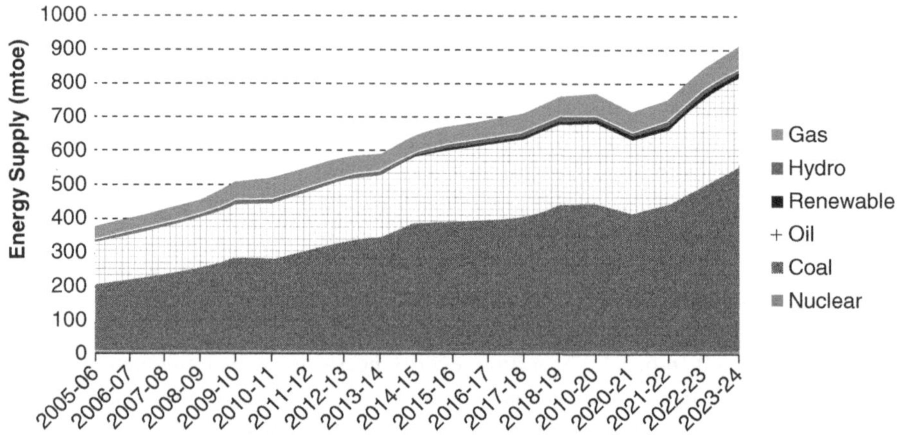

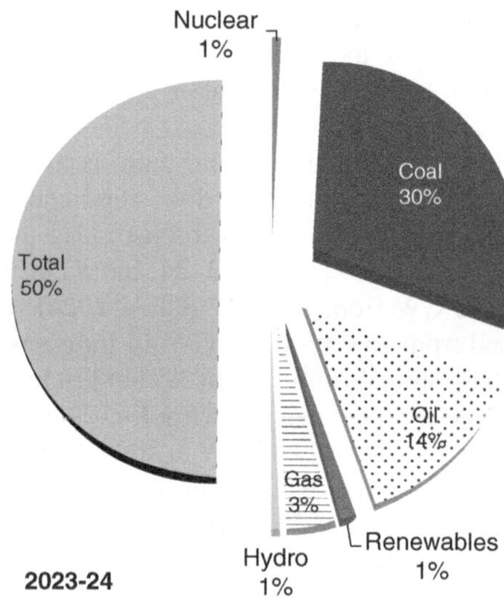

FIGURE 3.2 (a, b) India's energy mix 2023–24.

The technical losses occurred when energy dispersed in the conductors, apparatus utilized for transmission line, transformer at power exchange, sub-distribution line and magnetic losses in transformers before the electricity reached the consumers. In India, the technical losses are estimated at nearly 22.5% at normal level, and it is affected by the T&D network characteristics and the mode of operation. It is also observed that there are only about 30% of the total losses that occurred in transmission and sub-transmission lines, while most of the losses occurred in primary and secondary distribution lines. Therefore, to reduce or keep within limits of such type of losses could be achieved through proper planning and implementation of the primary and secondary distribution systems. There are technical losses that include some permanent/fixed losses that occur in the form of noise or heat but don't vary with increasing or decreasing the amount of the current; while some technical losses increase with the increasing amount of the current during electricity distribution.

The Non-Technical/Commercial Losses occur due to defective meter, inappropriate meter reading, and error in meter reading led to inaccurate billing of customer' energy consumption, lack of will power of administration, financial constraints related to effective monitoring, and problem of unmetered supply of energy as well as energy thefts.

URBANIZATION AND ENERGY INTENSIVE LIFESTYLE

Energy consumption in the residential sector accounts for 30% of the total final consumption in India, the second highest after the industrial sector (IEA, 2020a, 2020b; 2023). Residential energy consumption accounted for 82% of the total energy consumption while commercial energy consumption only imparts 18% in the residential and commercial sector excluding industrial and transport sector related to urban centres. The residential sector's energy consumption primarily includes the electricity requirement for lighting, cooking, HVAC (heating, ventilation, air conditioning). Other than electricity consumption, the urban residents also used liquefied petroleum gas (LPG); while kerosene, firewood, and biomass are still popular in rural residential energy consumption.

Due to the absence of a residential energy consumption survey in India; it is quite difficult to obtain the estimation of energy consumption but there are few studies (NITI Aayog, 2016) that helped to estimate the trends in residential energy consumption and how different appliances contribute to that. It is quite evident that lighting still has the largest share in residential energy consumption but this share of lighting showing a declining trend might be due to increasing use of electricity intensive appliances like refrigerators, water heaters, air coolers and air-conditioners. It is evident that about 90% of the total electricity consumption includes the energy consumption by lighting, fans, air-coolers, air-conditioners, refrigerators, water-heaters and televisions. Nowadays especially in the summer, space cooling proved to be the second highest consuming end-use segment in REC which accounts for the more use of fans, air-coolers and air-conditioners, and its share substantially increasing.

There are a number of factors which led to an increase in the total REC such as number of appliances owned by households, type and efficiency of those appliances (energy intensive or not), and usage of appliances. According to the 2011 census and other surveys, appliance ownership in Indian residents found to be still low as compared to major economies of the world, but growing fast. The fans are the most usable appliance with about 70% of the households own this, while it is followed by TV with 56% of the residents having this. Along with fans, other energy intensive appliances like refrigerators and air-conditioners were found to be quite low with about 23% and 2% respectively. The total REC determined by the resident's behaviour how they utilize the electrical appliances that also affected by two things: (a) the duration of usage, (b) the time of the usage in a day; based on the total duration of use of appliances might be useful to predict the total electricity consumed by the residential sector.

The residential sector behaviour also aligned with the increase in average floor area of residential buildings, reaching 64.5 square meters in 2020–21 while an all-India average showing at 54.2 square metres (Mo SPI, 2023). Kumar et al. (2018) also reported that the commercial buildings also showed an increasing phenomenon of total floor area; estimated to be 1.1 billion square meters in 2017 with future predictions of 1.78 about billion square meters by 2027. Further the construction material used in the buildings has its own U-value (represents a material's thermal conductivity, indicating its ability to conduct heat, measured in $W/K{\cdot}m^3$) and Albedo value (refers to the reflectivity of a surface, expressed as a number between 0 and 1); that is decided by the heat absorption or reflection by the material through conduction, convection, and radiation; the property of material as heat insulator or reflector improved with decrease in the U-value and increase in Albedo value. Apart from that building designs played a crucial role, lighting, ventilation and cooling and superior quality construction material with low U-values led to reduction in residential energy consumption significantly.

URBAN INFRASTRUCTURE AND URBAN HEAT ISLANDS EFFECT

The amount of sun's energy reflected by a surface is called albedo; so dark colours have an albedo close to zero, and pale colours have an albedo close to 100%. On average, Earth (including land, oceans, and ice) has an albedo of 31%, which means that about a third of the solar energy that comes to Earth is reflected into space. Therefore, to maintain the earth's average temperature, it is necessary to maintain the original surface colour or characteristic of the earth's surface.

However, cities are typically densely populated and are often a concrete jungle covered in buildings and asphalt. On the other hand, suburbs and rural areas tend to have a lot more vegetation and are less populated. Structures such as buildings, roads, and other infrastructure absorb and re-emit more sunlight than natural landscapes such as forests and water bodies. Urban areas with highly concentrated structures and limited greenery become "islands" with higher temperatures compared to outlying areas. The science behind the urban heat island effect lies in the amount of heat absorbed on surfaces such as concrete and asphalt and retained. Because these surfaces absorb a ton of heat from sunlight (typically 50°C–90°C higher than the air temperature), they slowly release it. These heat pockets are referred to as "heat islands." Heat islands can be formed under various conditions, including day or night, in small or large cities, in suburban areas, in northern or southern climates, and in any season. Due to these heat island effects, cities have these problems such as increased energy consumption (Santamouris, 2020; Maxwell et al., 2018; Zamuda et al., 2018), air pollution and greenhouse gas emissions, compromised human health and comfort (Gamble et al., 2013; U.S. Climate Change Science Program, 2008; Vaidyanathan et al., 2020) and impaired water quality (Somers et al., 2016).

FOSSIL FUEL BASED TRANSPORTATION AND CARBON EMISSIONS

Transport accounts for a significant share of the total energy in demand in India, with only three other countries dedicating a larger share of their energy to transport – the USA, China, and Russia (IEA, 2016). While the growth of transport energy demand in the USA, China, and Russia is declining, India's transport demand is set to increase, given that it is currently a growing economy with low per capita energy consumption for transport (IEA, 2021). The demand for transport will continue to grow in parallel with accelerating economic growth, making the transport sector in India critical from an energy use and decarbonization perspective. Urbanization is also characterized by the large number of public, commercial and private vehicles used for travelling and transportation of goods that often cause traffic congestion; led to air pollution. Further there is availability and quality of transport infrastructure in the cities and between the cities and towns or suburban or rural areas also responsible for fate and persistence of air pollution. In comparison to developing countries, developed or industrialized countries have more access to public transport in many transportation options, while developing countries lack good public transport or they even aren't necessarily distributed fairly or with consideration for the environment. As urbanization increases, the demand for public transportation grows dramatically every year in the developing world, where an estimated 1 billion people lack access to good roads and convenient public transport. It is reported that only 51.6% of the urban people worldwide have easy access to public transit in 2022 with notable regional variations (Figure 3.3).

Air pollution proved to be one of the worst health risks to people especially urban areas worldwide which was responsible for over 8.1 million premature deaths in 2021 including 0.7 million fatalities among children that are linked to air pollution mainly caused by vehicular emissions (WHO, 2021). It is also reported that long exposure of

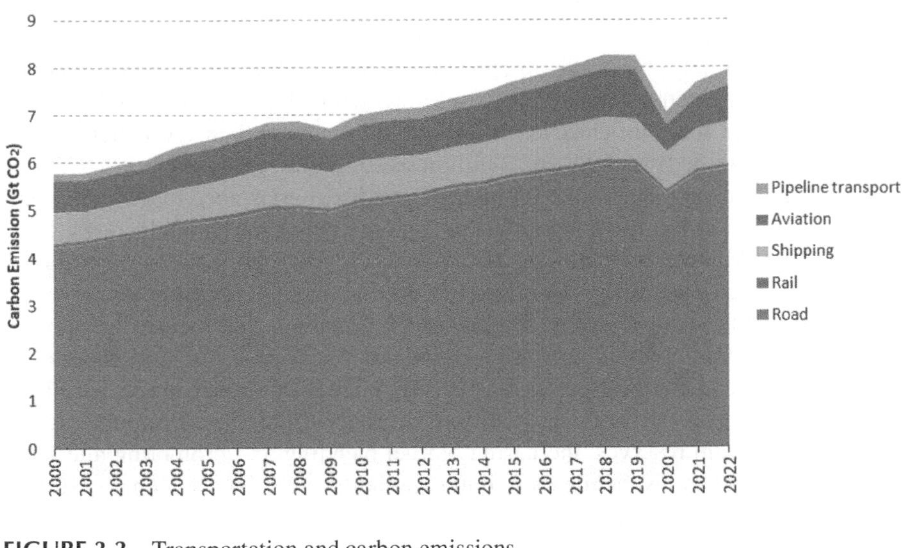

FIGURE 3.3 Transportation and carbon emissions.

traffic and associated air pollution also impact mental health, Alzheimer's disease, vision loss, and susceptibility to illnesses like COVID-19 (Carderón-Garcidueñas et al., 2020; Carrington, 2020; Farrow et al., 2020; Lu, 2020; Zang et al., 2022). In addition to various health issues and premature deaths, air pollution is thought to cost the world economy up to $2.9 trillion a year (3.3% of GDP) owing to emissions from fossil fuels alone. This alarming health catastrophe is further exacerbated by air pollution (CREA, 2020; HEI, 2024).

MOBILITY TRENDS AND FUTURE SCENARIO OF ELECTRIC VEHICLES

Due to increasing vehicles ownership and mobility platforms in India, there is increasing demand of petroleum products (diesel, gasoline, jet fuel and naphtha) from almost 2 million B/D (Barrels per day, a measure of oil output, represented by the number of barrels of oil produced in a single day) in 2023 to reach 7.1 million B/D by 2035 (S&P Global Commodity Insights). As India heavily depends on imports for their petroleum products, it has an effect on associated price volatility, impacts oil affordability, and leads to higher fiscal deficits resulting in long-term energy security. Having geopolitical issues with petroleum producers and risks related to supply choke points, India adopted a diversification strategy to reduce oil imports that involves ethanol blending with gasoline. India has already achieved a 20% average blend rate of ethanol in gasoline; to make this sustainable the government pushes for expanded feed-stocks for ethanol production, increasing ethanol production capacity like developing sugar mills for ethanol factories, lucrative ethanol prices and providing monetary assistance.

Electric vehicles sales reached nearly 14 million alone in 2023, in which 95% of the total sales were from China, Europe and US; and the total number electric cars also reached up to 40 million (IEA, 2023). India quite lags behind in the adoption of sales of electric cars but there is an increase of 70% year-on-year to 80,000 in electric car registrations and they make up the 10% growth rate for total car sales. It is evident that only 2% of the all cars sold in India; but in case of three wheelers and two wheeler electric vehicles the scenario is changed; they contributed over 50% and about 5% of the total electric vehicles in 2024. The Indian EV battery market is projected to surge from US$ 16.77 billion in 2023 to a remarkable US$ 27.70 billion by 2028. To supplement the EV market and fiscal assistance to the consumers, Production Linked Incentive (PLI) scheme and tax benefits to manufactures, Faster Adoption and Manufacturing of Electric Vehicles (FAME II) scheme to consumers, and the Go Electric campaign launched to fostering demand in recent years.

India's energy security is tied to the energy transition and focuses on adopting clean energy while continuing to use oil and gas for growth. India is undertaking various measures to address its increasing dependence on crude oil imports. These include investing in overseas upstream oil, diversifying crude oil imports, scaling strategic petroleum reserves, increasing biofuel blending and promoting alternative automotive power trains. While these initiatives address the issues of energy security and high oil import dependency to a degree, India needs to do more to successfully deal with its energy challenges.

OVERVIEW OF INDUSTRIAL ENERGY CONSUMPTION

The Indian Industrial sector was conventionally dependent on the electricity and fossil for their energy and heating and cooling needs. In industries, the energy needed to manufacture one unit of a product is known as specific energy consumption; further classified into specific electrical energy that is supplied from grid or captive power plants; and specific thermal energy primarily (utilized for direct heating processes) obtained from fossil fuels like coal. According to EMBER, 2024, India's heavy industries are still dependent on fossil fuel-derived thermal energy, which makes 89% of total industrial energy consumption in 2022; while 11% of the energy demand is supplemented through electricity. Further there is variation in electricity energy consumption across industries; the aluminium sector accounted for 80% of electricity consumption of the total energy consumption, followed by cement (10%), petrochemical (10%), steel (8.5%), and ammonia (4.8%). Although cement and steel sectors, still known for their higher total energy consumption due to their large production volumes (Table 3.1).

Due to some industries still dependent on energy derived from fossil fuels makes these sectors challenging to transition to a low-carbon model. Certain sectors use significantly more electricity per ton of output than others, making them prime targets for reducing carbon emissions. A substantial part of their energy usage can be swiftly switched to low-carbon sources through the use of renewable energy if suitable encouragement and legislation are established.

NET ZERO CARBON EMISSION GOAL AND USE OF RENEWABLE POWER

India's commitment towards adoption of clean energy alternatives and its pledge to become carbon neutral by 2070; to address increasing carbon emissions and climate change, India required to fulfil its half of electricity needs must be satisfied through renewable sources by 2030. Currently India holds fourth rank in renewable energy installed capacity in the world with 204.3 GW (Statista, 2025), fourth rank in solar power capacity with 105.65 GW, fourth in wind power capacity with 51.05 GW and good progress in small hydropower and biomass energy. Renewable

TABLE 3.1
Industry type and their total energy consumption

Industry type	Total energy consumption (in Mtoe)
Iron and steel	80
Cement	29
Fertilizer	20
Aluminium	17
Petrochemical	5
Chlor alkali	3
Textile	3
Pulp and paper	3

power generation capacity has expanded rapidly in recent years, with a compound annual growth rate of 15.4% between FY16 and FY23. As of FY23, India possesses a renewable energy capacity of 125.15 GW. India is expected to have the fastest growth in renewable electricity and by 2026; new renewable energy capacity additions are anticipated to double. Growing government backing and a better economic climate have made the sector more appealing to investors. With India aiming to meet its own energy needs, which are forecasted to total 15,820 TWh by 2040, renewable energy is poised to be a key player.

As of March 2025, renewable energy sources such as biomass, waste-to-power, and waste-to-energy collectively have a total installed capacity of 150.27 GW. As of October 2024, 44.72% of the total power installed capacity is from non-fossil-based sources. India's installed renewable energy capacity is forecasted to rise to approximately 170 GW by March 2025 from its December 2023 level of 135 GW, as per research agency ICRA. The country aims to have approximately 450 GW of renewable energy capacity installed by 2030, with solar energy anticipated to contribute around 280 GW (accounting for over 60%).

In the first three months of FY23, non-hydro renewable energy capacity expansion totalled 4.2 GW, compared to 2.6 GW in the first three months of FY22. Research conducted by the Council on Energy, Environment and Water's Centre for Energy Finance found that India's total installed power generation capacity reached 416 GW in FY23, with renewable energy accounting for 125 GW (30%) and hydroelectric power accounting for 47 GW (11%).

India's National Electricity Plan (Transmission) targets a 168 GW inter-regional capacity by 2031–32, involving the installation of 191,474 km of transmission lines and 1,274 GVA of transformation capacity. The system comprises 33.25 GW of HVDC links, cross-border interconnections and augmented private sector involvement to improve grid efficiency.

It reported that India installed 16.4 GW of solar capacity during January–September 2024, representing a 167% year-over-year growth. The Q3 FY24 witnessed a 78% increase in solar installations, reaching a total of 3.5 GW, a significant rise from the 2 GW seen in Q3 FY23. By September 2024, India's total installed solar capacity had reached 89.1 GW, with utility-scale projects making up more than 86% and rooftop solar covering almost 14%. India's total installed power capacity now contains around 20% solar power, while it makes up over 44% of its renewable energy capacity. The installed solar energy capacity has grown by 26 times in the last nine years and is at 73.32 GW as of December 2023. In 2023, India added 7.5 GW of solar power capacity. In January 2024, the added capacity from solar energy totalled 9,008.47 MW. At the end of December 2023, solar power comprised 16.9% of the total installed power capacity and 40.1% of the total installed renewable capacity. Renewable capacity's total share for solar power rose by 0.3% from the last quarter, at which time it represented 39.5% of the total capacity. In the initial eleven months of FY24, India has produced 75.57 BU of solar power.

India's renewable energy capacity is projected to reach 250 GW by 2026, as a result of a robust project pipeline. Renewable energy power generation excluding hydroelectric sources totalled 19.77 billion units in December 2024, an increase from the 16.76 billion units recorded in December 2023. Between April and December

2024, power generation from renewable energy sources totalled 189.48 billion units, a significant increase from 172.48 billion units during the same period in the previous year. Northern India is expected to become the hub for renewable energy in India, with a potential capacity of 363 GW and thanks to policies focused on the renewable energy sector.

REFERENCES

BEE. 2024. India energy scenario 2023–24. https://beeindia.gov.in/sites/default/files/BEE_India_Energy_Scenario_Report-2024.pdf

Calderón-Garcidueñas, L., Calderón-Garcidueñas, A., Torres-Jardón, R., Avila-Ramírez, J., Kulesza, R. and Angiulli, A. 2020. Neuroinflammation, hyperphosphorylated tau, diffuse amyloid plaques, and down-regulation of the cellular prion protein in air pollution–exposed children and young adults. *Journal of Alzheimer's Disease* 28(1), pp. 93–107.

Carrington, D. 2020. Air pollution linked to more severe mental illness. *The Guardian*. https://www.theguardian.com/environment/2021/aug/27/air-pollution-linked-to-more-severe-mental-illness-study

CEA (2024). National Electricity Plan. Volume II. https://powermin.gov.in/sites/default/files/uploads/National_Electricity_Plan_Volume_II_Transmission.pdf

CREA, 2020. Air quality improvements due to COVID 19 lock-down in India. https://energyandcleanair.org/air-quality-improvements-due-to-covid-19-lock-down-in-india/

Farrow, A., Miller, K.A. and Myllyvirta, L. 2020. Toxic air: The price of fossil fuels. https://policycommons.net/artifacts/2391360/the-price-of-fossil-fuels-full-report/3412789/

Gamble, J.L., Hurley, B.J., Schultz, P.A., Jaglom, W.S., Krishnan, N. and Harris, M. 2013. Climate change and older Americans: State of the science. *Environmental Health Perspectives* 121(1), pp. 15–22.

HEI. 2024. *State of global air 2024*. Health Effects Institute. https://www.stateofglobalair.org/resources/report/state-global-air-report-2024, https://niti.gov.in/sites/default/files/2020-01/IEA-India%202020-In-depth-EnergyPolicy_0

IEA 2016. World Energy OutloOk. Confirm. 2016, IEA, Paris, https://www.iea.org/reports/world-energy-outloOk. Confirm.-2016.

IEA. 2020a. India 2020: Energy policy review. https://iea.blob.core.windows.net/assets/2571ae38-c895-430e-8b62-bc19019c6807/India_2020_Energy_Policy_Review.pdf

IEA. 2020b. *World energy balance 2020*. International Energy Agency. https://www.iea.org/subscribe-todata-services/world-energy-balances-and-statistics

IEA. 2021. *India energy outlook 2021*. International Energy Agency. https://www.iea.org/reports/indiaenergy-outlook-2021.

IEA. 2023. Global EV OutloOk. Confirm. 2023, IEA, Paris https://www.iea.org/reports/global-ev-outloOk. Confirm.-2023, Licence: CC BY 4.0

Kumar, S., Sachar, S., Kachhawa, S., Goenka, A., Kasamsetty, S. and George, G. 2018. Demand analysis for cooling by sector in India in 2027. https://aeee.in/wp-content/uploads/2020/09/2018-Demand-Analysis-for-Cooling-by-Sector-in-India-in-2027-v2.pdf

Lu, J.G. 2020. Air pollution: A systematic review of its psychological, economic, and social effects. *Current Opinion in Psychology* 32, pp. 52–65.

Maxwell, K.J., Grambsch, A., Kosmal, A., Larson, L. and Sonti, N., 2018. Built environment, urban systems, and cities. In Reidmiller, D.R.; Avery, C.W.; Easterling, D.R.; Kunkel, K.E.; Lewis, K.L.M.; Maycock, T.K.; Stewart, B.C., eds. 2018. *Impacts, risks, and adaptation in the United States: Fourth national climate assessment* (Vol. II, pp. 438–478). US Global Change Research Program.

Mo SPI. 2023. Multiple indicator survey in India. NSS 78th Round (2020–21). https://www.mospi.gov.in/sites/default/files/publication_reports/MultipleIndicatorSurveyinIndiaf.pdf

NITI Aayog. 2016. India Energy Security Scenarios, 2047 user guide: Residential lighting and appliances. https://niti.gov.in/publication/annual-report

Prayas. 2016. Residential electricity consumption in India: What do we know? https://energy.prayaspune.org/images/pdf/residential_electricity_consumption_in_india.pdf

Ritchie, H. 2021. How have the world's energy sources changed over the last two centuries? https://ourworldindata.org/global-energy-200-years

Santamouris, M. 2020. Recent progress on urban overheating and heat island research. Integrated assessment of the energy, environmental, vulnerability and health impact. Synergies with the global climate change. *Energy and Buildings* 207, p. 109482.

Smil, V. 2017. *Energy transitions: Global and national perspectives. An imprint of ABC-CLIO*. LLC.

Somers, K.A., Berhnardt, E.S., McGlynn, B.L. and Urban, D.L. 2016. Downstream dissipation of storm flow heat pulses: A case study and its landscape-level implications. *Journal of the American Water Resources Association* 52(2), pp. 281–297.

Statista. 2025. *Primary energy consumption worldwide from 2000 to 2022, with a forecast until 2050, by energy source*. Statista Research Department. https://www.statista.com/statistics/222066/projected-global-energy-consumption-by-source

USCCSP. 2008. *Analyses of the effects of global change on human health and welfare and human systems*. U.S. Environmental Protection Agency.

Vaidyanathan. A., Malilay, J., Schramm, P. and Saha, S. 2020. Heat-related deaths—United States, 2004–2018. *Morbidity and Mortality Weekly Report* 69(24), pp. 729–734.

World Health Organization (2021). World Health Statistics 2021. https://www.who.int/publications/b/58471

Zamuda, C., Bilello, D.E., Conzelmann, G., Mecray, E., Satsangi, A., Tidwell, V. and Walker, B.J. 2018. Energy supply, delivery, and demand. In Reidmiller, D.R.; Avery, C.W.; Easterling, D.R.; Kunkel, K.E.; Lewis, K.L.M.; Maycock, T.K.; Stewart, B.C., eds. 2018. *Impacts, risks, and adaptation in the United States: Fourth national climate assessment* (Vol. II, pp. 174–201). US Global Change Research Program.

Zang, S.T., Luan, J., Li, L., Yu, H.X., Wu, Q.J., Chang, Q. and Zhao, Y.H. 2022. Ambient air pollution and COVID-19 risk: Evidence from 35 observational studies. *Environmental Research* 204, p. 112065.

4 Solar-powered urbanization

SOLAR POWER AND ITS CONVERSION

Solar energy, which is the radiant energy emitted by the sun and received by the earth, is considered an alternative energy source because of its benefits in terms of site selection and lower installation costs when compared to other types of renewable energy. Several approaches are being made to utilize the solar energy developed to tap into this energy source, including solar thermal (heating/cooling), solar power plants, photovoltaics (PV), solar architectural designs, and solar trees. There are several conversion approaches or methods that often implied as effective utilisation of the solar energy:

Passive Solar: Utilization of solar energy can be used in ventilation, heating and cooling the spaces without the aid of any mechanical or electrical appliances.

Solar thermal: Solar thermal systems convert solar energy into thermal energy by employing solar collectors or concentrating solar power technologies such as parabolic trough collectors.

Solar electricity: The electricity is generated by converting solar energy into electrical energy with the use of solid-state devices, such as solar PV technologies.

Hybrid thermal and electricity conversion: Converting solar energy to both electrical and thermal forms is achieved through a combination of solar collectors and solid-state devices, such as photovoltaic thermal collectors.

PASSIVE SOLAR ENERGY SYSTEMS

Solar passive heating within buildings captures sunlight, which is then released as heat during periods of darkness, such as at night. In the building design, south-facing glass apertures, i.e. windows or facades and thermal mass, i.e. concrete or bricks are often utilized for effective absorption, storage, and distribution of heat. Solar passive cooling uses passive ventilation shades and also thermal insulation to decrease the daytime heat and retain cool night-time air to regulate temperatures. In order to capture solar energy without the need for active systems, the building design should incorporate the following passive elements:

a. Properly oriented windows: Windows or other solar energy collection devices should usually be angled within 30° of south direction and must remain open or available through peak sunlight hours (9 a.m.–3 p.m.) during

DOI: 10.1201/9781003652229-4

the heating season, i.e. winter months that start by fall and ends with spring; while in the spring, autumn, and summer periods, the windows should be shaded to prevent overheating.

b. Thermal mass: In home building materials such as brick, stone, concrete, and tile known for their property to absorb warmth from sunlight during the winter (where indoor heating to be required also known as heating season). These materials also soak up the heat from the warm air that is generated indoors during the cooling season. Masonry materials like water and phase change products like paraffin wax, salt hydrates, and fatty acids store heat more efficiently, but they also function as both structural components and finish materials. Ensure that no objects obstruct sunlight from falling on thermal mass materials.
c. Distribution mechanisms: The heat from the sun is circulated from the absorption and storage points to various sections of the building via convection, conduction, and radiation. In certain buildings, compact fans, ducts and blowers assist in circulating warmth. Further light colour surfaces are known for releasing more heat than dark colour surfaces, and that can be utilized for selecting the thermal mass in passive solar homes.
d. Control strategies: Roof overhangs with appropriate dimensions can provide shade for south-facing windows in the summer months. Additional control methods consist of electronic sensors, including a differential thermostat that activates a fan; adjustable vents and dampers that manage heat transfer; low-emissivity curtains; movable insulating shutters; and awnings.

SOLAR THERMAL HEATING

Solar thermal heating utilizes the heat energy of the sunlight and transfers this to liquids, which is then utilized to heat water for domestic purposes and also be used to preheat the primary heating system. These solar heating systems can be categorized on the basis of required temperature of heating, encompassing low-temperature, and high-temperature applications.

- Low-temperature solar technologies, which operate below 70°C, include solar space heating, solar water heating, solar pond, and solar crop drying systems.
- Medium-temperature solar technologies, operating at medium temperatures (between 70°C and 200°C) include solar cooling, solar distillation, and solar cooking.
- High-temperature solar technologies operating above 200°C include high-temperature solar thermal power generation technologies like solar towers, parabolic troughs, and parabolic dishes (Table 4.1).

SOLAR THERMAL COOLING

Solar thermal cooling is categorized into four main classes, based on the cycles implied for the cooling process: (a) absorption type; (b) adsorption type; (c) desiccant

TABLE 4.1
Solar thermal heating technologies

Types	Solar collectors	Applications
Low-temperature technologies (operates below 70°C)	Flat Plate Collectors (FPCs), Evacuated Tube Collectors (ETCs)	Solar space heating, solar water heating, solar pond, and solar crop drying
Medium-temperature technologies (operates between 70°C–200°C)	Evacuated Tube Collectors with selective coatings, concentrated solar power (CSPs) like parabolic troughs for mid-temp	Solar cooling, solar distillation, and solar cooking
High-temperature technologies (operate above 200°C)	CSP systems like: Parabolic Trough, Solar Towers,	Solar tower, parabolic trough, and parabolic dish.

type; (d) ejector type. Firstly, the solar energy collected by the solar collectors, which is then converted to thermal energy to empower the cooling various cycles like absorption, adsorption, and desiccant cycles.

a. Absorption cooling systems: This technology is considered the oldest known cooling technology, but there is a difference in functioning, the refrigerator uses the conventional vapour compression system and relies on a mechanical compressor while a thermal compressor is employed in the absorption cooling process. The absorber in a thermal system holds a concentrated solution of coolant, whereas the generator contains a heat exchanger with a solution. In absorption chillers, the most commonly employed refrigerant-absorbent combinations are water/lithium bromide (H_2O/LiBr) and ammonia/water (NH_3/H_2O). In the water/lithium bromide pair, water acts as the refrigerant (coolant) and lithium bromide functions as the absorbent, whereas in the ammonia/water pair, ammonia is the refrigerant and water is the absorbent.
b. Adsorption cooling systems: The Adsorption cooling cycle is quite similar to the absorption cooling cycle, where liquid solution absorbs the refrigerant in absorption system, as in adsorption, the highly porous substance engages in the absorption of the refrigerant on its internal surface. The most commonly employed refrigerant-adsorbent combinations are activated carbon-methanol, carbon-ammonia, silica gel-water, activated carbon-ethanol, and zeolite-water. In the activated carbon-methanol pair, activated carbon acts as the solid sorbent and methanol acts as the refrigerant that adsorbs on the activated carbon.
c. Desiccant cooling systems: In desiccant systems, desiccant materials allow for the removal of air moisture content through the sorption process as in these materials. All desiccant materials have the ability to attract moisture with different capacities. For humid climates, the desiccant cooling system proved to be more efficient in achieving thermal cooling purposes. Additionally, the low temperature gain from solar energy can also be utilized by renewable energy, waste heat, and cogeneration to drive.

TABLE 4.2
Solar thermal cooling technologies

Types	Temperature	Solar collectors	Applications	Advantages
Absorption	80°C–110°C	Evacuated, parabolic	Large buildings, industry	Mature tech, efficient
Adsorption	60°C–90°C	Flat/evacuated tube	Remote areas, small scale	Fewer moving parts Lower efficiency
Desiccant (open)	50°C–80°C	Flat-plate/ evacuated tube	HVAC, dry climates	Good for humidity control, complex airflow system
Ejector	100°C–150°C	Parabolic trough	Remote or off-grid areas	Simple design, low maintenance, Low COP, less common

d. Ejector cooling systems: The solar ejector-based cooling system comprises three interconnected phases: solar phase, the ejector power phase, and the ejector cooling phase. The solar phase comprises a pump, solar collectors, and a heat exchanger (generator) to transfer heat to the ejector phase. In the solar phase, solar collectors heat the working fluid, typically water, to about 95°C. Next, the heated water flows out of the solar collectors to circulate through the generator. The generator transfers heat to the working fluid of the ejector power cycle without mixing the two distinct working fluid flows, resulting in a high-pressure and high-temperature vapour (Table 4.2).

SOLAR COLLECTORS

Solar collectors simply known as panels are broadly classified into: Flat Plate Collectors (FPC) and Evacuated Tube Collectors (ETC). Although it is cheap to use the FPCs, when it comes to domestic hot water the solar tubes system is more efficient.

a. Flat plate collectors: Flat plate collectors (FPCs) employ a dark flat-plate to absorb solar energy, covered with a transparent layer to minimize heat loss and permit the passage of solar energy, while heat from the absorber is extracted using a heat-transport fluid (air, antifreeze, or water) and a heat-insulating backing.
b. Evacuated tube collectors (ETCs) or solar tubes: They use a heat pipe in the core of the evacuated tubes, rather than passing liquid directly through them. These collectors comprise multiple evacuated glass tubes, which are fused with an absorber plate to a heat pipe. The heated pipe carries heat away from the hot end and transmits it to a fluid, usually propylene glycol, before distributing it to a domestic hot water tank heat exchanger and/or a hydronic space heating system. The vacuum surrounding the tube minimizes heat loss to the outside environment through both conduction and convection, resulting in higher efficiency than FPCs, particularly in colder climates.

CONCENTRATING SOLAR POWER

Concentrated solar power (CSP) systems are the mirrors and tracking systems used to capture and concentrate sunlight from a large area into a small, focused beam. The heat generated can be used for various purposes as a heat source, for example, in a traditional steam-driven power plant, in water desalination, or for preparing food. Of all the concentrating technologies, the most commonly used and well-developed ones are the parabolic trough and the solar power tower; the dish concentrator and linear Fresnel reflector, however, are less well-developed. Technologies for sun tracking and focused light are employed to enhance the system's efficiency. Thermal energy storage is typically employed in CSP systems to furnish the heat supply when sunlight is absent. Most CSP plants are equipped with storage capacity of up to 15 hours due to the efficient and cost-effective energy storage available in these systems.

PHOTOVOLTAICS OR SOLAR CELLS

When sunlight hits the solar cell, the photons in it are absorbed, causing the energy level of the electrons to rise, thereby freeing the electrons from their atomic cells. The presence of the p-n junction in the solar cell causes the electric field, which in turn forces the free electron into the n region. The positive charges are directed to the p region. The metal grid on the cell's surface bears the negative charge, whereas the positive charge is carried through the back plate (Figure 4.1) (Mark Hammonds, 1998). The phenomenon in question is referred to as the PV effect, with the semiconductor device converting solar energy into electricity being termed a PV cell.

Solar cells can be grouped together in series or parallel configurations to reach the desired power level (Ayoub et al., 2009; Ibrahim et al., 2009). The power generated by a solar cell is uncertain and entirely reliant on meteorological factors such as the sun's position, the amount of irradiation it receives, and ambient temperature (Ikegami et al., 2001; De Soto et al., 2006; Prajapati and Fernandez, 2020a, 2020b, 2020c, 2020d). A PV system or solar PV system is designed to provide usable power from the sun using PV technology. The PV system's modular nature makes available a wide range of capacities, from a few kilowatts for smaller roof-top or building-integrated systems, to hundreds of megawatts for larger utility-scale power stations. The majority of PV systems are now connected to the grid, whereas standalone systems constitute a relatively small segment of the market (Prajapati and Fernandez, 2021a, 2021b) (Table 4.3).

SOLAR POWER GENERATION IN INDIA

Considering abundance of solar irradiation throughout the year, India proved to be an ideal location to produce solar energy. With an average of 300 sunny days annually, this country presents an ideal setting for generating solar energy. India has the potential to produce up to 750 GW of solar energy, as stated by the National Institute of Solar Energy, a quantity that far exceeds the country's energy requirements. Rajasthan, Gujarat, and parts of Maharashtra are among the regions in the country with a high level of solar irradiation. India also has a significant amount of

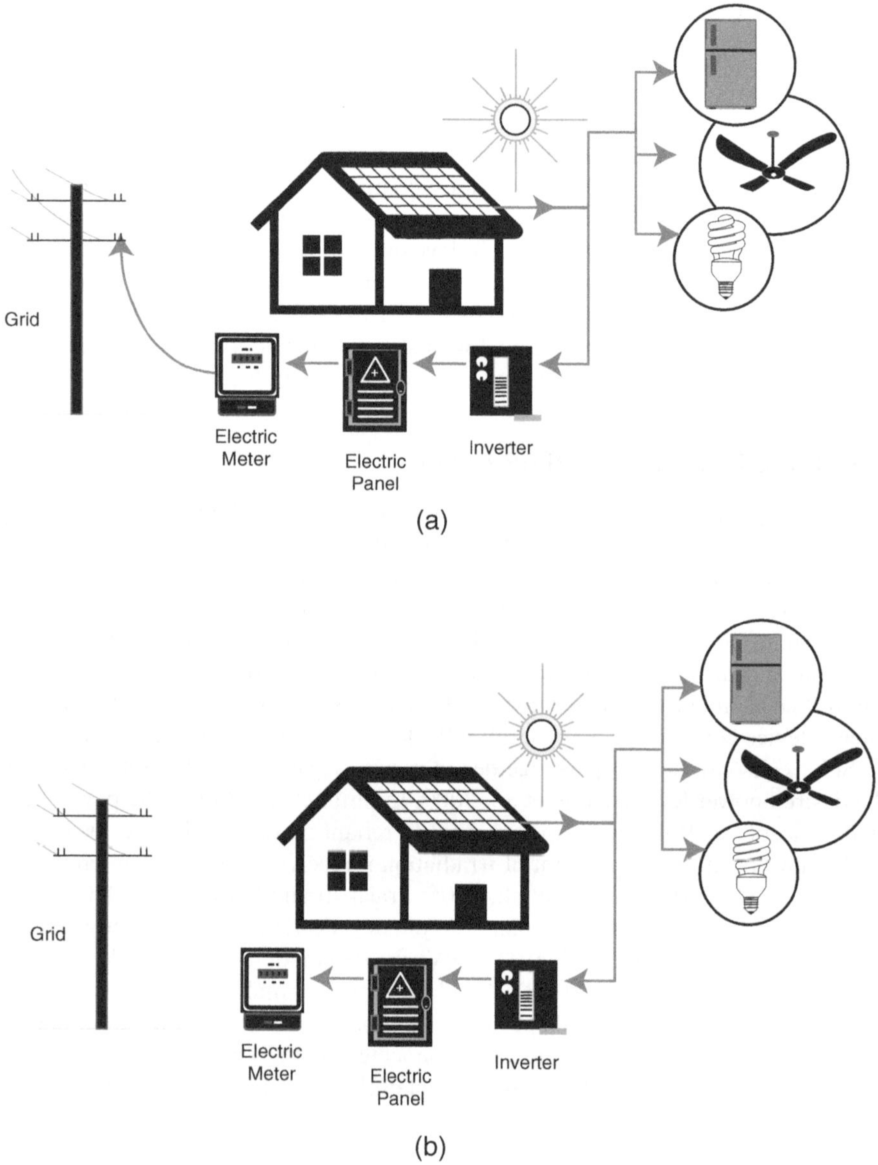

FIGURE 4.1 Solar-powered urbanization and its components.

land, especially barren lands that are suitable for solar power plants, with Rajasthan, Gujarat, and Tamil Nadu being particularly well suited for solar energy production.

India's solar energy journey took a major leap forward with the launch of the Jawaharlal Nehru National Solar Mission in 2010. The programme established a goal

TABLE 4.3
Type of photovoltaic cells

Type	Material	Efficiency (%)	Advantages
Monocrystalline Silicon (Mono-Si)	Pure single-crystal silicon	20–24	High efficiency, compact size, long lifespan, good in low light
Polycrystalline Silicon (Poly-Si)	Multi-crystal silicon	15–18	Cost-effective, simpler to produce
Amorphous Silicon (a-Si)	Non-crystalline silicon	6–10	Flexible, lightweight, performs well in diffuse light
Cadmium Telluride (CdTe)	Cadmium and tellurium compounds	10–18	Low production cost, good for hot climates
Copper Indium Gallium Selenide	Copper, indium, gallium, selenium	12–20	High efficiency among thin-film, flexible, lightweight
Perovskite Solar Cells	Hybrid organic-inorganic perovskite materials	20–26 (lab)	Rapid efficiency gains, low-cost materials, easy processing
Organic PV	Organic polymers or small molecules	5–12	Flexible, printable, lightweight, eco-friendly
Quantum Dot Solar Cells	Semiconductor nanocrystals (e.g., CdSe, PbS)	~10–15 (lab)	Tunable properties, potential for flexible and low-cost designs
Multi-Junction Solar Cells	Multiple semiconductor layers (e.g., GaAs, InGaP)	35–45 (lab)	Highest efficiency, ideal for satellites and concentrated solar power

of reaching 20 GW of solar energy by 2022. Encouraged by the sector's potential and the country's targets for renewable energy, this target was soon increased. India's green energy strategy rapidly incorporated solar energy as a crucial factor, backed by a robust policy framework and a business-friendly environment. Currently, solar power capacity accounts for roughly 90 GW, or nearly 45%, of India's total renewable capacity, demonstrating the country's significant advancements in green energy.

SOLAR PV SYSTEMS IN INDIA

Grid-connected PV system

Renewable-energy applications are increasingly driving the popularity of on-grid or grid-connected systems. Renewable energy sources, including solar PV systems, can be integrated into the electric power grid. This technology eliminates the need for energy storage and enables the return of excess power into the grid. A range of topologies and techniques have been developed to enhance the performance and efficiency of grid-connected systems. PV systems connected to the grid with battery storage systems.

Grid-connected PV with battery backup

The PV-battery power systems can offer battery backup during unusual grid situations and enhance power quality. Implementing battery energy storage systems in these configurations enables load leveling and power factor correction (Pathak et al., 2024). In these systems, the control of the voltage source inverters can ensure power equilibrium and enhance power quality under non-linear loads in a specific area (Mshkil and Alwan, 2022), and the BESSs can also meet grid code and fault ride-through specifications by employing a dual control strategy that alternates between constant battery current control and continuous direct current (DC)-link voltage control (Berger et al., 2021). Incorporating BESSs into grid-connected systems can also offer supplementary services such as frequency stabilization, improving dependability and producing income (Wang et al., 2022; Fu et al., 2023).

Off-grid system

Off-grid systems provide a solution for areas with limited access to a reliable energy supply from the public power grid. These systems employ solar panels to convert sunlight into DC power, which is then stored in batteries. A charge controller controls the charging and discharging operations of the battery pack, and an inverter converts the stored DC power into alternating current (AC) for appliances and other AC loads (Fu et al., 2023). Off-grid solar power systems are cost-effective and efficient for providing electricity to areas without grid access (Pratheeba et al., 2023). Furthermore, off-grid solar PV systems can produce electricity in isolated regions, rendering local generators unnecessary (Kaur et al., 2015). A hybrid system that combines vertical wind turbines and solar cells can be utilized to guarantee uninterrupted energy generation in areas without a grid connection (Jahan et al., 2022).

Solar-powered urbanization: concept and its dimensions

Solar-powered urbanization is an integrated approach for using solar power for meeting the increasing energy demands of the urban sector that includes residential and commercial, industrial and agriculture and food processing sectors. But there are several sub-sectors: (a) Lighting; (b) HVAC (Heating, Ventilation and Air conditioning); (c) Water supply and water recycling; (d) Transportation; (e) Agriculture especially urban farming; that are integral part of all the sectors from low to high consumption and also varies or might be specific for a particular sector. These sub-sectors have a rapidly increasing energy consumption profile and will require more energy consumption over the years that need to be addressed to reduce the carbon emissions, mitigate the climate change and ensure energy security.

Solar-powered lighting provides several advantages for both the environment and public health. The widespread use of kerosene lamps, still prevalent in many regions, releases several pollutants, such as black carbon, closely associated with climate change and a primary cause of global warming, equivalent to 240 Mt of CO2 emissions annually. The use of kerosene lamps is problematic due to potential fire hazards and their toxicity, primarily due to high levels of harmful particulates (Mills, 2016). Meanwhile, the increasing adoption of renewable energy sources has increased demand for home storage systems for electricity generated by PV installations as costs decrease. These systems store excess energy generated from solar or wind sources when it is produced sporadically, making it available for household appliances when necessary. More and more private residences are incorporating storage systems to briefly store electricity produced from PV systems; instead of sourcing power from the grid, these households can make use of renewable energy to power their home electronics and appliances. Considerable progress is being made in smart lighting systems. Solar lighting fixtures can now be automated, controlled, and monitored remotely thanks to the integration of Internet of Things (IoT) technologies. The use of smart sensors, including light, motion, and temperature detectors, notably increases the adaptability of lighting. This innovation not only enhances convenience for consumers but also increases energy efficiency further.

Conventional heating, ventilation, and air conditioning systems heavily rely on electricity produced from fossil fuels, which significantly contributes to carbon emissions. In contrast, solar-powered heating and cooling systems utilize the inexhaustible energy of the sun, converting sunlight into electricity via PV panels. This renewable energy source lessens our reliance on non-renewable resources. This also minimizes environmental harm, making it a desirable choice for eco-friendly cooling and heating alternatives. Solar-powered HVAC systems consist of solar panels, inverters, batteries, and HVAC equipment. Solar panels, usually installed on rooftops or exposed areas, harness sunlight and convert it into DC electricity. DC converters are often used to convert DC power into the AC current that the building's electrical system requires. Excess energy can be stored in batteries for future use or fed back into the grid via net metering schemes. Integrating solar power with heating, ventilation, and air conditioning systems minimizes energy consumption. Heat pumps, a key component of these systems, harness the sun's energy to either heat or cool air, thus lessening the burden on traditional heating and cooling systems. Smart thermostats and controls also boost efficiency by adjusting the temperature according to real-time data, reducing energy waste even further.

Renewable energy-dependent water management systems are significantly impacted by solar panels. Implementing their application in water treatment facilities, irrigation systems, and pumping systems, in addition to energy-efficient solar desalination methods, fosters sustainable practices in water management and tackles problems associated with water scarcity. Advances in solar-powered technologies increasingly provide a beneficial pathway for the efficient and environmentally sustainable management of water resources. Clean water is becoming a priority, and solar-powered water treatment facilities are emerging as a viable solution to meet the growing demand. These facilities reduce their reliance on conventional energy sources by tapping into solar power, resulting in more environmentally sustainable water treatment systems. Integrating solar panels into systems for water and irrigation pumping enhances the management of water resources by making it more sustainable and efficient. Researchers led by Vishnupriyan et al. in 2022 conducted experiments to find the optimal tilt angle and orientation for maximizing solar energy use in water pumping systems in remote locations. Energy-efficient solar desalination technologies have revolutionized sustainable methods for desalinating water. These pioneering solutions address freshwater scarcity and offer more eco-friendly alternatives to conventional desalination methods that use solar power.

The future of transportation is expected to be electric vehicles (EVs) due to increasing costs of conventional fuel for petrol and diesel-based vehicles. As a result, the energy demand will be greater when EVs are introduced into the public transportation system. This review chapter assesses a solar-powered charging station for EVs by examining how tilting the solar panel at various angles affects the maximum efficiency and power it can obtain from sunlight, taking into account different wavelengths of solar radiation. Solar panels, specifically PV ones, can be used to charge EVs in a sustainable manner. Harnessing the solar potential in office buildings for EV charging on the job, and capitalizing on the long parking duration at work (Prajapati and Fernandez, 2019), paves the way for the deployment of vehicle-to-grid technology. This chapter compares and researches various system architectures for an EV-PV charger. Electric cars are gaining widespread acceptance in numerous countries worldwide. EVs are demonstrating greater energy efficiency and a reduced environmental impact compared to internal combustion engine vehicles. The limited availability of charging stations is a major barrier to the widespread adoption of EVs globally. As EV usage continues to increase, more public areas are installing EV charging stations. On the other hand, if EVs are charged via an existing utility grid that is powered by a fossil fuel-based generation system, then it impacts the distribution system and cannot be environmentally friendly. Generating electricity from solar-powered PV panels offers tremendous potential, making charging EVs from these panels a fantastic solution and a sustainable step towards environmental conservation.

Solar energy, a plentiful and adaptable source, is becoming increasingly popular across diverse sectors including residential homes, industrial, manufacturing, agriculture, and technology. This technology can help conserve water resources, decrease reliance on the grid, lower electricity expenses, and produce extra revenue. Although it has been used in agriculture for a long time, there is a growing need for its use in a broader range of functions. The agricultural sector is confronted with various difficulties including water shortages, rising energy expenditures, and the requirement for environmentally friendly practices (Mahmood and Al-Ansari, 2024), and as a result, solar energy offers a versatile and dependable solution. Faced with global climate change and escalating energy requirements, the agricultural industry is now relying on renewable energy sources for environmentally sustainable alternatives ((Kumar et al., 2014). Solar energy has proven itself as a significant resource that enables farmers to lower expenses, boost productivity, and foster eco-friendly practices. This comprehensive guide will examine the different ways in which solar energy is revolutionizing agriculture, its advantages, and how it can be utilized to boost productivity in farming. The adoption of solar energy is increasingly being adopted in agriculture (Figure 4.2).

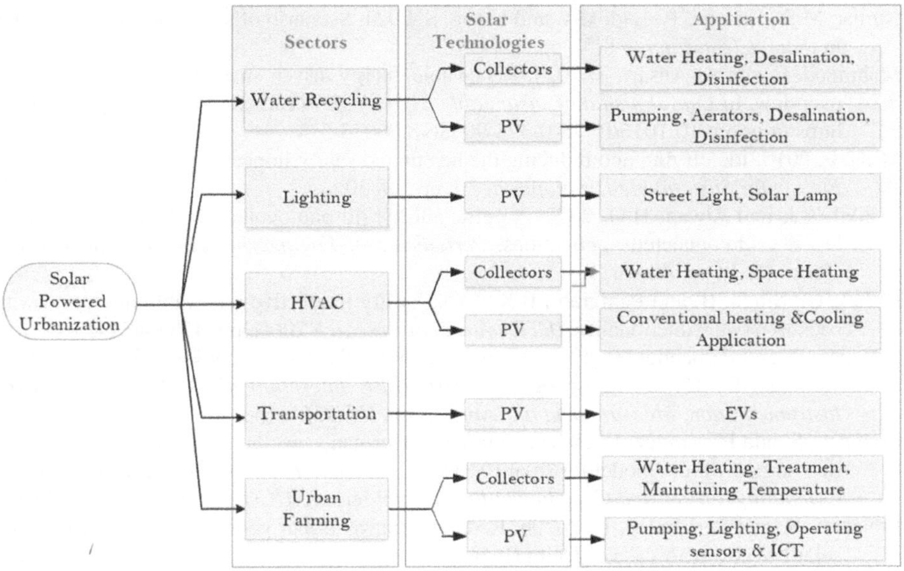

FIGURE 4.2 (a, b) Grid connected and off-grid PV systems.

REFERENCES

Ayoub, N., Batres, R. and Naka, Y. 2009. An approach to wicked problems in environmental policy making. *WSAES Transactions on Environment and Development* 5(3), pp. 229–239.

Berger, M., Kocar, I., Farantatos, E. and Haddadi, A. 2021. Dual control strategy for grid-tied battery energy storage systems to comply with emerging grid codes and fault ride through requirements. *Journal of Modern Power Systems and Clean Energy* 10(4), pp. 977–988.

De Soto, W., Klein, S.A. and Beckman, W.A. 2006. Improvement and validation of a model for photovoltaic array performance. *Solar Energy* 80(1), pp. 78–88.

Fu, Y., Li, W., Xiong, N., Fang, Y., He, X. and Wang, Z. 2023. Influence of off-grid/grid-connected operation on stability of large-scale photovoltaic system. *Energy Reports* 9, pp. 904–910.

Hammonds, M. 1998. *Getting power from the sun*. Chemistry and Industry.

Ibrahim, M., Sopian, K., Daud, W.R.W., Alghoul, M.A., Yahya, M., Sulaiman, M.Y. and Zaharim, A. 2009. Solar chemical heat pump drying system for tropical region. *WSEAS Transactions on Environment and Development* 5(5), pp. 404–413.

Ikegami, T., Maezono, T., Nakanishi, F., Yamagata, Y. and Ebihara, K., 2001. Estimation of equivalent circuit parameters of PV module and its application to optimal operation of PV system. *Solar Energy Materials and Solar Cells* 67(1–4), pp. 389–395.

Jahan, I.S., Blazek, V., Misak, S., Snasel, V. and Prokop, L. 2022. Forecasting of power quality parameters based on meteorological data in small-scale household off-grid systems. *Energies* 15(14), p. 5251.

Kaur, P., Jain, S. and Jhunjhunwala, A. 2015, June. Solar-DC deployment experience in off-grid and near off-grid homes: Economics, technology and policy analysis. In *2015 IEEE First International Conference on DC Microgrids (ICDCM)* (pp. 26–31). IEEE.

Kumar, M.S., Kumari, P., Lad, Y.A. and Maria, S. 2024. Scenario of solar energy and policies in India. *Clean Energy* 8(5), pp. 117–128.

Mahmood, F. and Al-Ansari, T. 2024. Renewable energy driven sustainable greenhouse: An overview. In *Encyclopedia of sustainable technologies* (2nd ed., Vol. 1–4, pp. V2–652). https://doi.org/10.1016/B978-0-323-90386-8.00073-5

Mills, E. 2016. Identifying and reducing the health and safety impacts of fuel-based lighting. *Energy for Sustainable Development* 30; pp. 39–50.

Mshkil, A.J. and Alwan, H.O. 2022. An efficient hybrid photovoltaic battery power system based grid-connected applications. *Periodicals of Engineering and Natural Sciences (PEN)* 10(3), pp. 409–421.

Pathak, G., Singh, B. and Panigrahi, B.K. 2024. Utility interactive renewable microgrid with backup power functionality. *IETE Journal of Research* 70(4), pp. 4333–4344.

Prajapati, S. and Fernandez, E. 2019, August. Rooftop solar PV system for commercial office buildings for EV charging load. In *2019 IEEE International Conference on Smart Instrumentation, Measurement and Application (ICSIMA)* (pp. 1–5). IEEE.

Prajapati, S. and Fernandez, E. 2020a, October. Performance evaluation of membership function on fuzzy logic model for solar PV array. In *2020 IEEE International Conference on Computing, Power and Communication Technologies (GUCON)* (pp. 609–613). IEEE.

Prajapati, S. and Fernandez, E. 2020b. Reducing conventional generation dependency with solar PV power: A case of Delhi (India). *Energy Sources, Part A: Recovery, Utilization, and Environmental Effects* 47(1), pp. 3750–3761. https://doi.org/10.1080/15567036.2020.1859006.

Prajapati, S. and Fernandez, E. 2020c. Relative comparison of standalone renewable energy system battery storage requirements for residential, industrial and commercial loads. *International Journal of Renewable Energy Technology* 11(2), pp. 111–125.

Prajapati, S. and Fernandez, E. 2020d. Solar PV parking lots to maximize charge operator profit for EV charging with minimum grid power purchase. *Energy Sources, Part A: Recovery, Utilization, and Environmental Effects* 47(1), pp. 3319–3329. https://doi.org/10.1080/15567036.2020.1851325.

Prajapati, S. and Fernandez, E., 2021a. Capacity credit estimation for solar PV installations in conventional generation: Impacts with and without battery storage. *Energy Sources, Part A: Recovery, Utilization, and Environmental Effects* 43(22), pp. 2947–2959.

Prajapati, S. and Fernandez, E. 2021b. Standalone solar PV system for grey water recycling along with electric load for domestic application. *International Journal of Sustainable Engineering* 14(5), pp. 933–940.

Pratheeba, C., Muthuvinayagam, M., Ramkumar, M.S., Bhat, C.R., Maniraj, P. and Kumar, N.S. 2023, May. A review of an off grid solar DC system for rural houses. In *2023 7th International Conference on Intelligent Computing and Control Systems (ICICCS)* (pp. 1837–1843). IEEE.

Wang, K., Liu, S., Kopsidas, K. and Forsyth, A. 2022, June. Benefits of grid-connected battery energy storage systems for reliability enhancement. In *2022 17th International Conference on Probabilistic Methods Applied to Power Systems (PMAPS)* (pp. 1–6). IEEE.

5 Solar-powered lighting

DAYLIGHTING AND ITS IMPORTANCE

Daylighting is the phenomenon of intentional use of visible light including direct sunlight and diffused-skylight within buildings to reduce reliance on artificial lighting and also reduce electricity usage. By establishing a connection to the changing and constantly shifting trends of outdoor light, daylighting fosters an engaging, eco-friendly and productive environment for residents, while lowering the overall energy consumption up to one-third of the total energy consumption by the buildings (Edwards and Torcellini, 2002). A daylighting system is composed not only of sources of daylight like the windows and skylights, but also includes a lighting control system that responds with the daylight levels. This system can decrease electric lighting energy consumption when sufficient ambient light from daylight is available. Additionally, the arrangement of windows in a building, known as fenestration, should be designed in a way that minimizes direct sunlight to reach the place of work or don't bother the occupants' visual space. Alternatively, suitable glare-minimizing solutions like blinds or shades should be attainable (DiLaura et al., 2011).

Daylighting serves as a crucial aspect in contemporary architecture, allowing natural light to enter a space through openings in the building, thus replacing or reducing the need for artificial lighting. This approach is not only helpful in lowering the building's energy consumption but also improves the visual comfort for the residents (Chen et al., 2014a, 2014b). Utilizing daylight has been considered as an effective method for elevating the visual quality of interior space and enhancing energy efficiency in the buildings. Manning (2006), Alrubaih et al. (2013) suggested that implementing daylighting could lead to a decrease of 223 Mt of $CO2$ emissions or a reduction of 24 GW in energy demand. Nonetheless, excess exposure to daylight may result in blazing, warming issues, and thermal discomfort for the building's residents. Xue et al. (2014) indicates that the quality of daylighting influences the luminous comfort experienced by those inside the building.

INTEGRATION OF DAYLIGHTING IN BUILDING DESIGN

Natural light in a building undoubtedly improves the energy efficiency, leading to financial and energy savings but it also requires lighting control solutions or photo sensors that dim or turn off artificial lighting in the condition of adequate daylight. For this, there are different lighting control mechanisms that include manual, time based, and automatic control systems which were explored. Although assessing the energy savings from lighting controls and their payback periods can be challenging (Bodart and De Herde, 2002; Doulos et al., 2008; Ihm et al., 2009), the possible energy reduction from various daylight-responsive dimming systems can be helpful in reliable projection through use of digital simulations and confirmed through field

DOI: 10.1201/9781003652229-5

data (Doulos et al., 2008; Ihm et al., 2009; Li et al., 2014). Research has shown a decrease in artificial lighting energy consumption within building spaces due to the use of lighting controls (Ihm et al., 2009).

Nonetheless, the absence of simplified evaluation tools that can provide insights into the effectiveness and cost-efficiency of daylighting remains a significant reason for the hesitance of building professionals to integrate daylighting solutions into their designs. However, there is a shortage of simplified assessment tools that can offer insights into the practicality and cost-effectiveness of daylighting can still be regarded as one of the primary factors why construction professionals hesitate to include daylighting elements in their designs (Ihm et al., 2009).

Various daylighting technologies have been created to enhance the performance of natural lighting in indoor spaces of buildings. Numerous researchers are focused on developing daylighting design strategies and technologies for buildings. They have explored both design and technological methods, including layout and parametric modifications to building designs, as well as the implementation of daylight harvesting technologies (Wong, 2017). These devices, referred to as daylighting systems, are positioned near or at the openings of building envelopes, allowing them to reflect and redirect incoming natural light into interiors, thereby improving lighting conditions. Two prevalent forms of daylighting systems include side-lighting and top-lighting (Karlen and Spangler, 2023). A typical example of side-lighting is the conventional vertical window opening, whereas top-lighting is exemplified by openings in the roof or ceiling structures of buildings. Natural light can be introduced into a building through standard vertical windows, clerestory windows, or roof lights, along with various remote distribution systems. With standard vertical windows, light intensity diminishes rapidly as one moves further from the window. The deeper a room is, the less well it is illuminated by daylight (Apian-Bennewitz et al., 1998). Designs incorporating top-lighting require careful examination to prevent potential overheating due to improper top-lighting designs. To address this issue, innovative daylighting

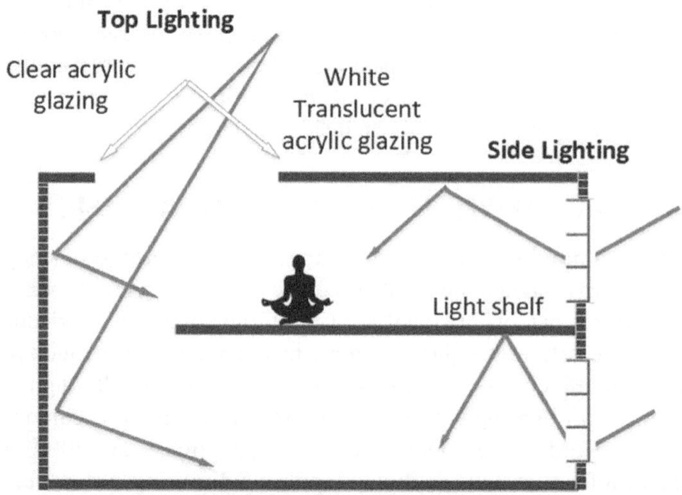

FIGURE 5.1 Top lighting and side lighting.

systems featuring new components and technologies have been created to channel sunlight further into building interiors while minimizing overheating concerns (Sick, 1997). Littlefair (1989, 1996) described innovative daylighting systems that can be categorized into two groups: light guiding systems and light transporting systems. Both varieties of daylighting systems can be utilized to enhance daylight distribution in building spaces and manage direct sunlight. Despite earlier studies aimed at improving natural illumination in buildings, particularly in deep floor plan structures with limited daylight access, the focus has primarily been on office buildings rather than industrial structures, where lighting represents a significant electrical expense (Chen et al., 2014a, 2014b) (Figure 5.1).

EVOLUTION OF LIGHTING SYSTEMS

Throughout history, the quest for improved and brighter sources of light has been a constant endeavour since fire was first discovered (Linhart and Scartezzini, 2011). The Romans later lit the exteriors of their homes with oil lamps, employing specific servants to manage these light sources. In 1417, a law instituted by the Mayor of London mandated that all households display a lantern in front during winter months (Tregenza and Wilson, 2013). During the 1700s, light fixtures consisted mainly of inadequately constructed oil lamps and candles, which produced unpleasant odours and excessive smoke. It was not until the 1800s that gas lighting began to replace both candles and oil lamps. In 1809, Humphry Davy invented the carbon arc lamp for the first time in history, which provided quite luminous light but needed a battery (earliest battery) to power up while these batteries were known for their quick discharge (Ref). The first patent for an incandescent lamp was granted to Frederick de Moleyns in 1841; his lamp utilized glass and a charcoal filament within a partial vacuum, though the charcoal caused the lamp to become dirty and dim over time.

The late 1800s witnessed important advancements in artificial lighting. A patent for an incandescent lamp featuring a carbon filament was awarded in 1874 to Henry Woodward and Mathew Evens. Eventually, Woodward and Evens sold their patent to Thomas Edison, who, along with his team, experimented with various substances to invent a comparatively effective carbonized filament and later observed that a lamp with a carbonized bamboo filament was capable of lasting up to 1,200 hours, leading Edison to secure his second incandescent lamp patent. Although Edison is not credited with inventing the first light bulb, his "Edison screw-in lamp" remains the standard for household lighting to this day. In 1911, William D. Coolidge developed tungsten filament lamps, which were subsequently modified by Irving Langmuir in 1913. Langmuir experimented tungsten filament by twisting the filament and supplemented the glass case with an inert gas to enhance the performance of the lamp. Up until 1925, all lamps were constructed from clear glass. Marvin Pipkin introduced a method to minimize glare by adding a silica coating to the inside, resulting in the creation of the first frosted incandescent lamps. The late 1950s marked the beginning of the "halogen era," with scientists at GE introducing the first halogen lamp. Although these lamps proved to have a longer lifespan, they required careful handling. In 1976, Edward Hammer developed the first spiral compact fluorescent lamp (CFL).

As is often the case with innovative products, the initial discovery of the LED happened by chance. While working on a laser diode, Gary Pittman and James Biard stumbled upon the first "light emitting diode" (LED) in 1961. The earliest LED lights found use as indicators and display lights during the early 1970s, with many researchers continuing to advance LED technology to create diodes in various colours. A crucial element was still necessary to produce a bright white LED, which was achieved in 1993 by Shuji Nakamura through the invention of the first bright blue emitting diode. Today, these diodes are combined with different phosphor materials to create various Kelvin temperatures in lamps that are currently manufactured. The implications for climate change would be profound and transitioning away from outdated incandescent bulbs and shifting to better and improved lighting solutions like CFLs or LEDs represents a straightforward and cost-effective method to significantly cut carbon emissions (Huang, 2004; Huang et al., 2006).

SOLAR-POWERED LIGHTING

The lighting sector is responsible for using about 15% of electricity consumption and contributes roughly 5% of carbon emissions at world level. The lighting utilized a substantial amount of electricity which led to significant carbon emissions. Wong (2017) found out that artificial lighting can account for one third of the electricity expenses or could reach up to 40% of annual energy consumption for a residential building and 20–30% of the total energy consumption in commercial establishments, or 35% of the overall electricity consumption in standard office premises. In the constructed establishments, solar energy is utilized in various aspects, including lighting and heating. The concept and utilization of passive solar energy in the building's designs, employs architectural components to collect, store, and distribute solar energy, is gaining significance. Some of the passive solar energy design strategies, such as daylighting and space heating, proved to be among the most direct and efficient methods.

As the use of renewable energy continues to grow, home energy storage systems linked to photovoltaic (PV) setups are becoming more attractive due to decreasing prices. These systems store excess energy produced during intermittent solar or wind generation and provide it when needed for household appliances. An increasing number of homes are adopting storage solutions to temporarily hold electricity generated by their photovoltaic systems; instead of relying on the grid, households can now use renewable energy to power their devices. In 2024, smart lighting technologies are making considerable progress. By incorporating Internet of Things (IoT) capabilities, solar lighting fixtures can now be automated, remotely controlled, and monitored. The integration of smart sensors—for light, motion, and temperature detection—significantly enhances the adaptability of lighting.

Solar-powered lighting offers numerous benefits for both the environment and public health. The burning of kerosene lamps, which remain common in various areas, emits numerous pollutants like black carbon, closely linked to climate change and responsible for global warming, resulting in an equivalent of 240 Mt of CO2 emissions each year. The use of kerosene lamps also presents risks due to potential fire hazards and their toxicity, as they contain high levels of harmful particulates (Mills, 2016).

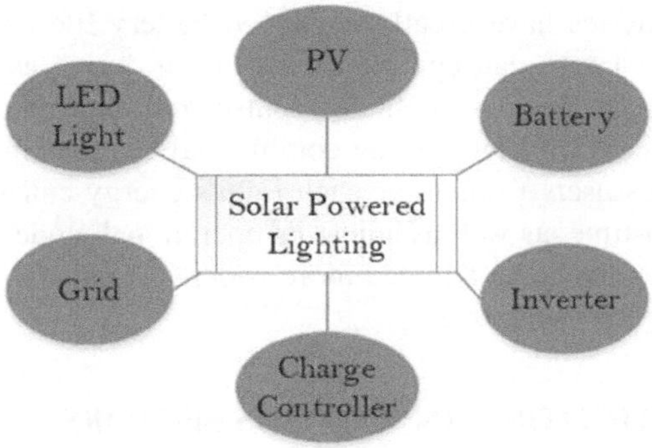

FIGURE 5.2 Solar-powered lighting.

Conversely, the rising adoption of renewable energy sources has boosted the appeal of home storage systems for electricity generated by photovoltaic (PV) installations, as their costs decline. These systems store surplus energy generated from solar or wind sources when it's irregularly produced, making it available for household appliances when needed. An increasing number of private homes are integrating storage systems to temporarily hold power generated from photovoltaic systems; rather than drawing power from the grid, these households can utilize renewable energy to operate their home devices and appliances. Currently smart lighting systems are making considerable advancements. With the integration of Internet of Things (IoT) technologies, solar lighting fixtures can now be automated, controlled, and monitored remotely. Utilizing smart sensors (such as light, motion, and temperature detectors) significantly enhances the versatility of lighting. This innovation not only increases convenience for consumers but also further enhances energy efficiency (Figure 5.2).

SOLAR PANEL TECHNOLOGIES

The efficiency of monocrystalline silicon solar panels has experienced remarkable improvements. Advanced photovoltaic technology has elevated the energy conversion efficiency of these solar panels to new heights, allowing solar lamps to recharge their batteries in a shorter time frame. This progress is vital for areas that receive minimal sunlight or encounter difficult weather conditions. The effectiveness of monocrystalline silicon solar panels has greatly increased. Groundbreaking photovoltaic technology has raised the energy conversion efficiency of solar panels to unprecedented levels, enabling solar lamps to charge their batteries in a relatively brief time. This innovation is particularly significant in regions with scarce sunlight or unfavourable weather.

ENERGY STORAGE

Revolution in Energy Storage Solutions: The effectiveness of solar lighting systems is based on progress in energy storage technology. Advances in lithium-ion and lithium iron phosphate batteries have greatly enhanced battery life and operational time. Consequently, solar lamps can operate reliably even during extended rainy spells. A unified management system facilitates centralized oversight of numerous solar lights, making it perfect for large-scale commercial or municipal lighting projects. This system enables users to monitor each light's energy consumption and operational status in real-time, as well as adjust its operational mode from a distance. By 2024, this technology is being adopted more widely around the world, especially in the context of smart city initiatives.

SMART SOLAR LIGHTING IN VARIOUS SECTORS

a. Remote and Off-Grid Lighting: The use of solar lamps is especially crucial in distant areas and regions without electricity. With advancements in energy storage technology and the efficiency of solar panels, solar lighting options in these locations have become more reliable. Smart luminaries can automatically adjust their brightness, conserving energy while ensuring that basic lighting is maintained even when battery levels are low.
b. Residential and Small Business Use: Smart solar wall lights and garden fixtures have become popular among residential customers and smaller businesses. These products combine stylish design with smart capabilities, providing practical lighting solutions for home patios, gardens, or small retail locations.
c. Commercial and Municipal Lighting: Smart solar lighting solutions present considerable potential in both commercial and municipal sectors. The lighting needs of urban areas, including streets, parking lots, and public spaces, often require significant power resources. Intelligent solar lights can fulfill these demands by employing efficient solar panels and energy storage systems while reducing operational costs through remote management via a centralized platform.

ROOFTOP SOLAR PV SYSTEM

In these systems, solar panels can be installed on the rooftops of residential, commercial, and industrial buildings (Prajapati and Fernandez, 2020). The electricity generated is used to meet the load demands. In some instances, this power can be used directly for non-critical loads without requiring battery storage, which is referred to as direct-coupled solar PV systems. A common use for these systems is in water pumping applications. This configuration is meant to pump water only during daylight hours. The volume of water pumped is entirely dependent on the sunlight exposure received by the PV panels and the type of pump used. As there is no need for storage, the system operates only when sunlight is present. Due to the straightforward setup of these systems, their costs are generally lower in comparison. Battery-coupled solar PV systems include a battery backup to ensure a continuous power supply when solar energy is not available. The costs for maintenance and replacement pertain to the battery storage (Prajapati and Fernandez, 2021).

BUILDING-INTEGRATED PHOTOVOLTAICS

The integration of solar photovoltaics or thermal technologies into building architecture and construction could transform the utilization of solar power for the lighting sector and minimize the carbon footprints for a sustainable and environmentally conscious future. To integrate solar panel installations into buildings safely and effectively, structural considerations are essential. Engineers meticulously assess load-bearing capacities and the compatibility of architectural elements to enhance solar panel positioning and boost energy collection. The efficiency and performance of solar energy in solar-powered structures illustrate the considerable benefits derived from solar panel integration. Building-integrated photovoltaics (BIPV) systems can be crafted to harmonize with conventional building materials and designs, or they can be employed to achieve a more modern look. The following examples illustrate how PV modules can serve as appealing features of building exteriors. Photovoltaics can be integrated into various components of building envelopes, including.

FACADES

Solar cells can enhance or substitute for standard view windows or spandrel glass. Although these installations are positioned on vertical surfaces, which lower the intensity of solar insulation, the overall area of a facade can help mitigate the decreased power per unit area. Based on different integrated PV functions, facade BIPVs can be categorized into two types: BIPV cladding and curtain walls, both of which form essential parts of the facade structure. Therefore, it is crucial to take into account the fundamental characteristics of the building envelope, such as weatherproofing and waterproofing. Additionally, when designing the latter type of wall, one must consider indoor visibility and direct sunlight alongside the facade. It is important to highlight that prior studies have concentrated on the merging of BIPV cladding walls and phase-change materials (PCMs) to enhance the efficiency and heat dissipation of PV systems. Research has shown that in different regions, BIPV systems incorporated with PCMs can keep the surface temperature of the PV below 29° for a specific duration (130 minutes).

ROOFING

The incorporation of PV into roofing solutions can effectively replace batten and seam metal roofing, traditional three-tab asphalt shingles, and ceramic tiles. It is essential to remember that such installations necessitate sufficient ventilation to maintain lower cell temperatures. Unlike non-integrated PV roof systems (like building-attached photovoltaic (BAPV) systems), roof BIPV systems blend existing roofing materials, such as tiles, into the structure without requiring extra mounting frameworks like racks and rails. BIPV tiles can resemble traditional tiles in terms of colour and size, catering to the needs of sensitive architectural contexts.

SKYLIGHTS

Implementing PV in skylight systems can represent a cost-effective application of PV and a compelling design element. Similar to PV windows, the semi-transparency

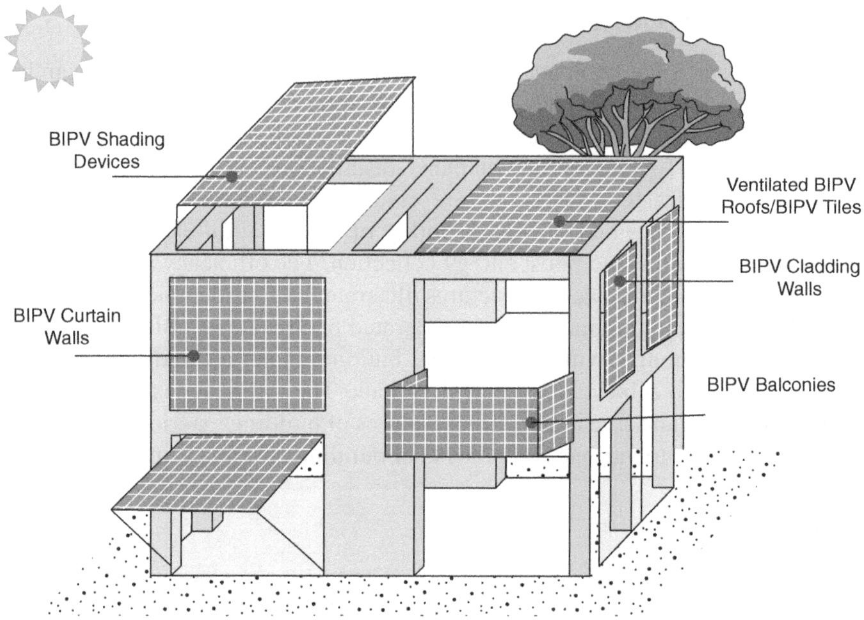

FIGURE 5.3 Integration of PVs into building architecture.

allows for visual connections with the outside environment while providing diffused natural light.

AWNINGS

Photovoltaics can be integrated into awnings or slightly inclined, saw-tooth canopy designs. Semi-transparent modules deliver filtered sunlight below while offering additional architectural advantages like passive shading (Figure 5.3).

REFERENCES

Alrubaih, M.S., Zain, M.F.M., Alghoul, M.A., Ibrahim, N.L.N., Shameri, M.A. and Elayeb, O. 2013. Research and development on daylighting in offices: A review. *Renewable and Sustainable Energy Reviews* 21, pp. 94–111.

Apian-Bennewitz, P., Goller, M., Herkel, S., Kovach-Hebling, A. and Wienold, J. 1998. Computer-oriented building design: Advances in daylighting and thermal simulation tools. *Renewable Energy* 14(1–4), pp. 351–356.

Bodart, M. and De Herde, A. 2002. Global energy savings in offices buildings by the use of daylighting. *Energy and Buildings* 34(5), pp. 421–429.

Chen, Y., Liu, J., Pei, J., Cao, X., Chen, Q. and Jiang, Y. 2014a. Experimental and simulation study on the performance of daylighting in an industrial building and its energy saving potential. *Energy and Buildings* 73, pp. 184–191.

Chen, Y., Ng, E., An, X., Ren, C. and Lee, M. 2014b. Developing a design support tool for sunlight access analysis in high-density urban environments. *Building and Environment* 71, pp. 17–28.

DiLaura, D.L., Houser, K.W., Mistrick, R.G. and Steffy, G.R. 2011. *The lighting handbook: Reference and application* (10th ed.). Illuminating Engineering Society of North America.

Doulos, L., Tsangrassoulis, A. and Topalis, F. 2008. Quantifying energy savings in daylight responsive systems: The role of dimming electronic ballasts. *Energy and Buildings* 40(1), pp. 36–50.

Edwards, L. and Torcellini, P. 2002. *A literature review of the effects of natural light on building occupants* (NREL/TP-550-30769). National Renewable Energy Laboratory. https://www.nrel.gov/docs/fy02osti/30769.pdf

Huang, B.J. 2004. Development of a low-cost LHP for commercial application. In *13th International Heat Pipe Conference*, Shanghai, China. Sept. 21–25, 2004.

Huang, B.J., Wu, M.S. and Wu, C.J. 2006, August. Development and field test of a long-lasting solar LED lighting system. In *Word Renewable Energy Congress IX* (p. 590).

Ihm, P., Nemri, A. and Krarti, M. 2009. Estimation of lighting energy savings from daylighting. *Building and Environment* 44(3), pp. 509–514.

Karlen, M. and Spangler, C. 2023. *Lighting design basics*. John Wiley & Sons.

Li, D.H., Cheung, A.C., Chow, S.K. and Lee, E.W. 2014. Study of daylight data and lighting energy savings for atrium corridors with lighting dimming controls. *Energy and Buildings* 72, pp. 457–464.

Linhart, F. and Scartezzini, J.L., 2011. Evening office lighting–visual comfort vs. energy efficiency vs. performance?. *Building and Environment* 46(5), pp. 981–989.

Littlefair, P. 1989. *Innovative daylighting systems*. Information paper 22/89. IHS BRE Press.

Littlefair, P. 1996. *Designing with innovative daylighting*. IHS BRE Press.

Manning, W.J. 2006. Plants in urban ecosystems: Essential role of urban forests in urban metabolism and sustainability. *Environmental Pollution* 142(3), pp. 389–395.

Mills, E. 2016. Identifying and reducing the health and safety impacts of fuel-based lighting. *Energy for Sustainable Development* 30, pp. 39–50.

Prajapati, S., and Fernandez, E. 2020. Reducing conventional generation dependency with solar Pv power: A case of Delhi (India). Energy Sources, Part A: Recovery, Utilization, and Environmental Effects, 47(1), pp. 3750–3761. https://doi.org/10.1080/15567036.2020.1859006

Prajapati, S. and Fernandez, E., 2021. Capacity credit estimation for solar PV installations in conventional generation: iImpacts with and without battery storage. Energy Sources, Part A: Recovery, Utilization, and Environmental Effects 43(22), pp. 2947–2959.

Sick, F. 1997. Tageslichtnutzung (Daylighting). In Marko, A., Braun, P. (eds), *Thermische Solarenergienutzung an Gebäuden* (pp. 123–152). Springer Berlin Heidelberg.

Tregenza, P. and Wilson, M. 2013. *Daylighting: Architecture and lighting design*. Routledge.

Wong, L. 2017. A review of daylighting design and implementation in buildings. *Renewable and Sustainable Energy Reviews* 74, pp. 959–968.

Xue, P., Mak, C.M. and Cheung, H. 2014. The effects of daylighting and human behavior on luminous comfort in residential buildings. *Energy and Buildings* 81, pp. 47–56.

6 Solar-powered heating, ventilation, and air conditioning

NATURAL HEATING, VENTILATION AND AIR CONDITIONING (HVAC)

Traditional architectural systems like Vastu Architecture of India emphasize the use of natural elements—orientation, materials, airflow, sunlight, and spatial planning—to regulate indoor climate without mechanical systems (Jain and Gautam, 2018; Vastu Shastra, 2003). Vastu Architecture has Cardinal Directions based orientation as core principle: (a) North considered ideal for study rooms and work areas that receives diffused, cool light in India and other tropical regions. (b) East is considered ideal for main entrances and living spaces. Morning sunlight offers natural heating and disinfection, (c) South direction: for storage or staircases to buffer heat. In cold climates, it can store warmth; (d) West direction: Avoid as main living areas; use for less occupied rooms to reduce heat gain in the evening.

Further construction materials can be used for passive temperature control like (a) use of thick mud, lime, or stone walls which provide excellent insulation led to keep interiors cool in summer and warm in winter; (b) Natural roofing materials like clay tiles or terracotta which allow heat dissipation; (c) Wooden jaalis and chik blinds that filter sunlight and facilitate airflow. Further the application of the high thermal mass materials (like stone or concrete) to absorb heat during the daytime and release it at night. Traditional homes used step-wells, tanks, and ponds in the northeast (Ishan kona) for cooling and spiritual cleansing. Small water features like fountains in courtyards help drop ambient temperature via evaporation. Green landscaping, especially near air inlets, cools incoming air (Bansal et al., 1994; Jain and Garg, 2017).

It also focuses on the concept of cross ventilation and natural air circulation, which involves aligning windows and doors on opposite walls to create cross-ventilation, latticed screen) or clerestory windows high on walls to release hot air, use courtyards such as central open space that cools surrounding rooms by convection; and build wind tunnels such as narrow corridors that funnel cool air from shaded regions or water features. Vastu recommends designing to capture early morning light and avoid harsh afternoon sun. Deep overhangs, verandas, and pergolas: Reduce solar gain while allowing light. Angled louvres and ventilated facades: Let in breezes while shading interiors. Design night ventilation systems with openable skylights or roof vents to flush out warm air (Bansal et al., 1994; Jain and Garg, 2017; Jain and Gautam, 2018).

DOI: 10.1201/9781003652229-6

CONVENTIONAL HVAC AND ENERGY CONSUMPTION

Since the earliest home building in history, humans have tried various methods such as the use of fires, ice, and windows for regulating the indoor environment for survival and thermal comfort. Later coal played a crucial role in heating the inside of homes especially in Europe (which also became the reason for many deaths due to pollution and breathing issues). But major and revolutionary changes have happened since the first air conditioner was developed in the early 20th century; later they became the essential part of the homes and buildings.

In the current scenario, HVAC systems widely used for a controllable indoor environment around the urban world, leading to a surging energy demand by the residential and commercial buildings. U.S. Energy Information Administration (USEIA) (2021) estimated that HVAC systems alone are responsible for 40% energy in residential buildings and 36% in commercial buildings and also responsible for significant carbon emissions. In the old days, HVAC systems were considered luxury resources but nowadays have become almost an essential part of the urban lifestyle that demands more thermal comfort. However, the need for HVAC systems is also aligned with the climate, global warming and also the income of the residents. The richest nations like the US and EU have higher HVAC consumption rates compared to emerging economies (such as China and India), whereas Russia exhibits the highest percentage due to its colder climate. In conclusion, HVAC usage accounts for approximately 12% of global final energy consumption and can reach around 25% in affluent or cold regions, like the EU and Russia (Ürge-Vorsatz et al., 2015; IEA, 2022).

Traditional HVAC systems regulate temperature and ventilation throughout an entire indoor space based on a set schedule, fixed temperature goals, and predetermined ventilation rates, failing to take into account the actual occupancy levels. This leads to energy waste during unoccupied periods and dissatisfaction with indoor environmental quality, which includes thermal comfort and perceived air quality. Therefore, it is crucial for policies to target this significant energy-demanding end-use, especially in developed areas, through enhancements and updates to building envelopes and solar-powered HVAC systems (Pérez-Lombard et al., 2012).

SOLAR-POWERED HVAC

Solar-powered HVAC are the systems that can be utilized to directly heat and cool buildings using thermal energy or to produce electricity PV systems. These systems provide both heating and cooling and also function as such hybrid systems that fully operate on solar power by optimizing their capacity to utilize available solar power, or prioritize solar energy when available, or supplementing any shortfall with grid power. These solar heating and cooling systems come with environmental advantages due to lower carbon emissions and offer improved indoor environmental quality by minimizing pollutants and eliminating combustion exhaust. As severe weather events become more frequent, relying on independent onsite solar energy, rather than outdated utility grids, is a wise choice for enhancing energy efficiency, saving costs, and ensuring business continuity (Kong et al., 2022; Niaki et al., 2023) (Figure 6.1).

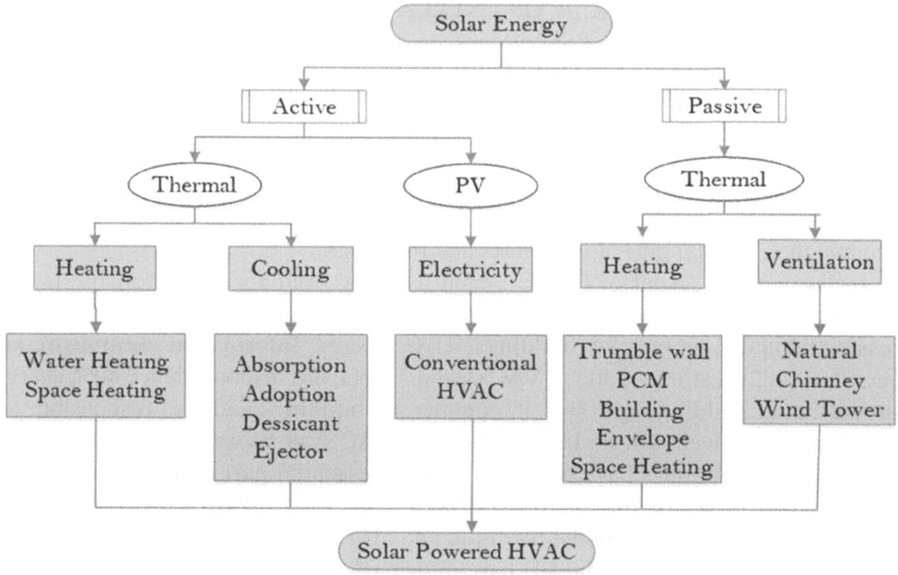

FIGURE 6.1 Solar-powered HVAC.

SOLAR SPACE HEATING SYSTEMS

The heating, ventilation, and cooling processes in the buildings are considered to be interconnected and work together to maintain flow of fresh air, temperature regulation, and thermal comfort. To facilitate effective heating or cooling to a building, natural ventilation and air conditioning are the integral parts, which can be achieved through various natural or passive approaches such as natural ventilation, Trombe wall, solar chimney, wind towers, phase changing material (PCM), thermal insulation, etc.

TROMBE WALL

Trombe wall (named after engineer Felix Trombe), considered a passive solar building heating approach, a south facing wall (north facing in the southern hemisphere) made up of dark coloured material that might be concrete, covered by glass to create air vent between wall and the glass (U.S. Department of Energy, 2022). During the daytime, these walls heat up from the solar radiation to warm up the air that goes upwards and it is consequently replaced by the cool air from building through the bottom opening of the wall; resulting in the entry of warm air into the building through the top opening of the wall. This process continuously takes place to warm up the indoors of the building indoors, but the vents at top and bottom are closed during night-time to avoid the reverse process of warm air escaping to the vent and cool air enters in the building (Figure 6.2).

PHASE CHANGE MATERIALS (PCM)

PCMs are the construction materials that are known to absorb and release significant amounts of energy upon the changes in phase of these materials like solid to liquid

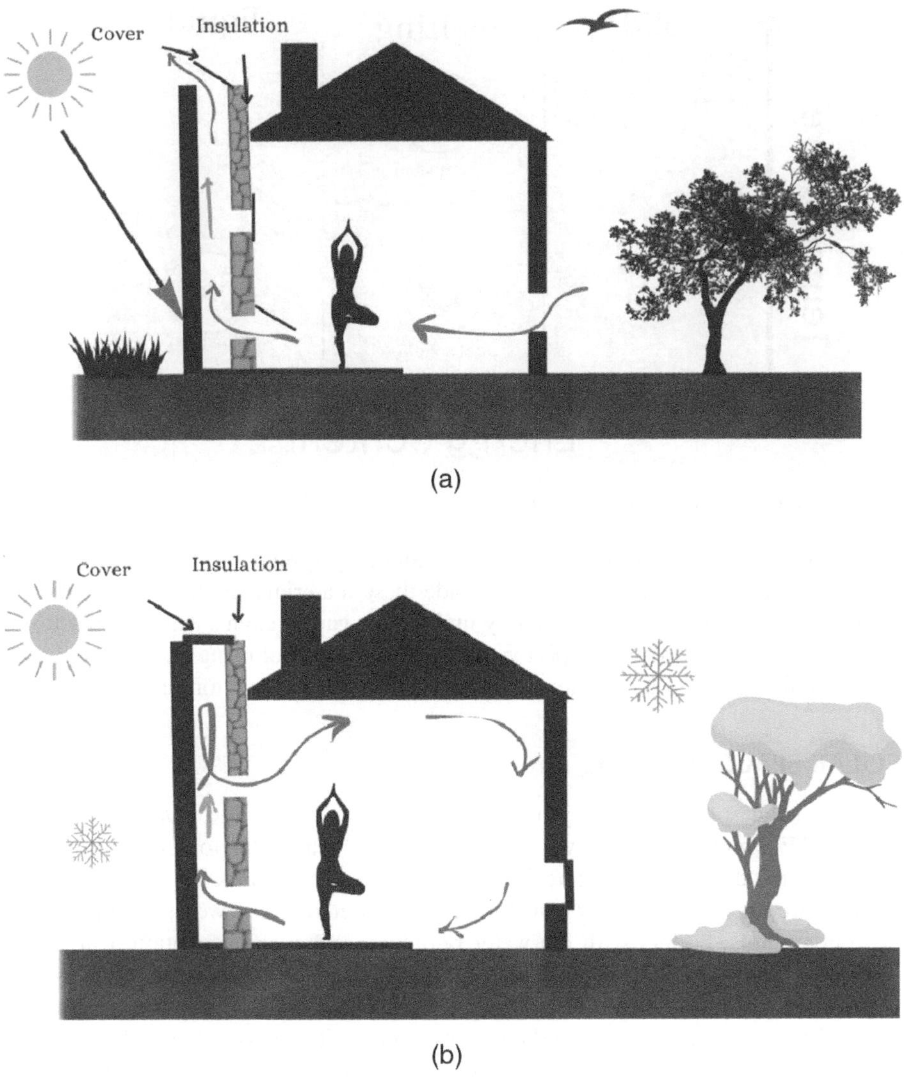

FIGURE 6.2 (a, b) Trombe wall.

or liquid to vapour. As the temperature increases, materials undergo phase change, leading to absorption and release of significant amounts of energy, also called latent heat storage (Rathore and Shukla, 2019). There are three types of phase change materials: Organic PCMs, Inorganic PCMs, and Eutectic PCMs; (a) Organic PCMs: paraffin and fatty acids; (b) Inorganic PCMs: Salt hydrates and metals; (c) Eutectic PCMs: Mixture of two or more salts (Figure 6.3).

BUILDING ENVELOPE

The outer surface of the building could be covered with a layer of such materials that could be used for thermal insulations either mineral fibre (rock wool, slag wool and

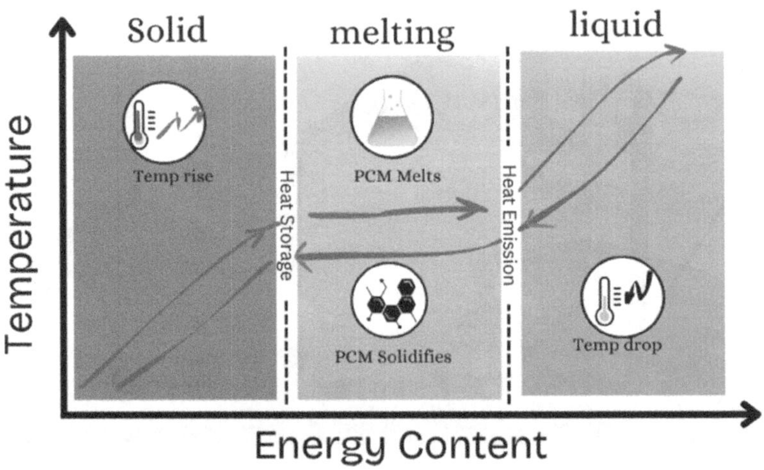

FIGURE 6.3 Phase change materials.

glass wool) or cellular plastic (polyurethane and polystyrene sheets or foam) or plant derived (cellulose, wool, cotton, flax). Beside these materials, another natural thermal insulator i.e. air that can be utilized by making air gap in cavity wall construction to increase the thermal insulation performance. This method of using air gap is only suitable for building in mild climate conditions but it is not suitable for temperate regions.

SOLAR SPACE HEATING

In active solar space-heating systems, solar energy is utilized to heat a heat transfer fluid (HTF which is primarily liquid or air) in a solar collector and then transfer the solar heat directly to the interior space or to a storage tank for later use. Liquid systems are more often used when storage is included, and are well suited for radiant heating systems, boilers with hot-water radiators, and even absorption heat pumps. Thus, the distribution fluid (hot water) that is used by existing heating systems (e.g. radiator, radiant floor) is circulated through a heat exchanger in the storage tank. As the hot-water passes through the heat exchanger, it is warmed, and then returned to the heating system. If the solar system cannot provide adequate space heating, an auxiliary or back-up system provides the additional heat.

Two types of control schemes are commonly used on solar collectors on building-scale applications: on-off and proportional. With an on-off controller, a decision is made to turn the circulating pumps on or off depending on whether or not useful output is available from the collectors. With a proportional controller, the pump speed is varied in an attempt to maintain a specified temperature level at the collector outlet.

SOLAR WATER HEATING

Solar water heating systems consist of solar collectors and storage tanks. There are two categories of solar water heating systems: active systems, which utilize circulating pumps and controls, and passive systems, which do not. The two types of solar water heating systems are: (a) Direct circulation systems, where pumps move household water through the collectors and into the residence; these are effective in regions with minimal freezing; (b) Indirect circulation systems, where pumps circulate a non-freezing heat-transfer fluid through the collectors and a heat exchanger, subsequently heating the water that flows into the home. These systems are favoured in areas susceptible to freezing conditions. Most solar water heaters need a well-insulated storage tank. Solar storage tanks feature an extra outlet and inlet linked to the collector. In two-tank configurations, the solar water heater raises the temperature of water before it reaches the conventional water heater. In one-tank setups, the backup heater is integrated with the solar storage in a singular tank. There are three types of solar collectors utilized for residential use.

FLAT PLATE COLLECTORS

Flat plate collectors employ a dark flat-plate to absorb solar energy, covered with a transparent layer to minimize heat loss and permit the passage of solar energy, while heat from the absorber is extracted using a heat-transport fluid (air, antifreeze, or water) and a heat-insulating backing.

EVACUATED TUBE COLLECTORS OR SOLAR TUBES

They use a heat pipe in the core of the evacuated tubes, rather than passing liquid directly through them. These collectors comprise multiple evacuated glass tubes, which are fused with an absorber plate to a heat pipe. The heated pipe carries heat away from the hot end and transmits it to a fluid, usually propylene glycol, before distributing it to a domestic hot water tank heat exchanger and/or a hydronic space heating system. The vacuum surrounding the tube minimizes heat loss to the outside environment through both conduction and convection, resulting in higher efficiency than flat-plate collectors (FPCs), particularly in colder climates.

INTEGRAL COLLECTOR STORAGE SYSTEMS

Integral Collector Storage systems referred to as ICS, are also known as "batch" or "bread box" water heating systems, where the collector and tank are integrated into a single unit, eliminating the need for circulation pumps and electronic controls, as the household tap water is heated and stored within this combined heat storage and collection device. ICS, or "batch" systems, are constructed using several large diameter tubes ranging from 4 to 8 inches (100–200 mm) in diameter, with one or more black-painted storage tanks located inside an insulated and weather-tight glazed box. Under normal water pressure, cold water enters the solar collector, which preheats and stores the water. The solar heat transfer from the collector to the water occurs through natural convection, requiring no external energy, thus making it a fully passive system.

SOLAR VENTILATION

Solar ventilation utilizes solar energy to improve airflow and ventilation within buildings, offering a sustainable and efficient alternative to traditional methods. It primarily involves harnessing solar radiation to preheat ventilation air or power fans, reducing reliance on electrical grids. This can be achieved through solar chimneys, roof-mounted solar panels, or integrated systems that combine solar energy with ventilation ducts. Here's a more detailed look at solar ventilation techniques.

NATURAL VENTILATION

The natural ventilation or induced ventilation is the natural process which occurs due to differences in temperature of the air parcels as the air parcels heat up leading to

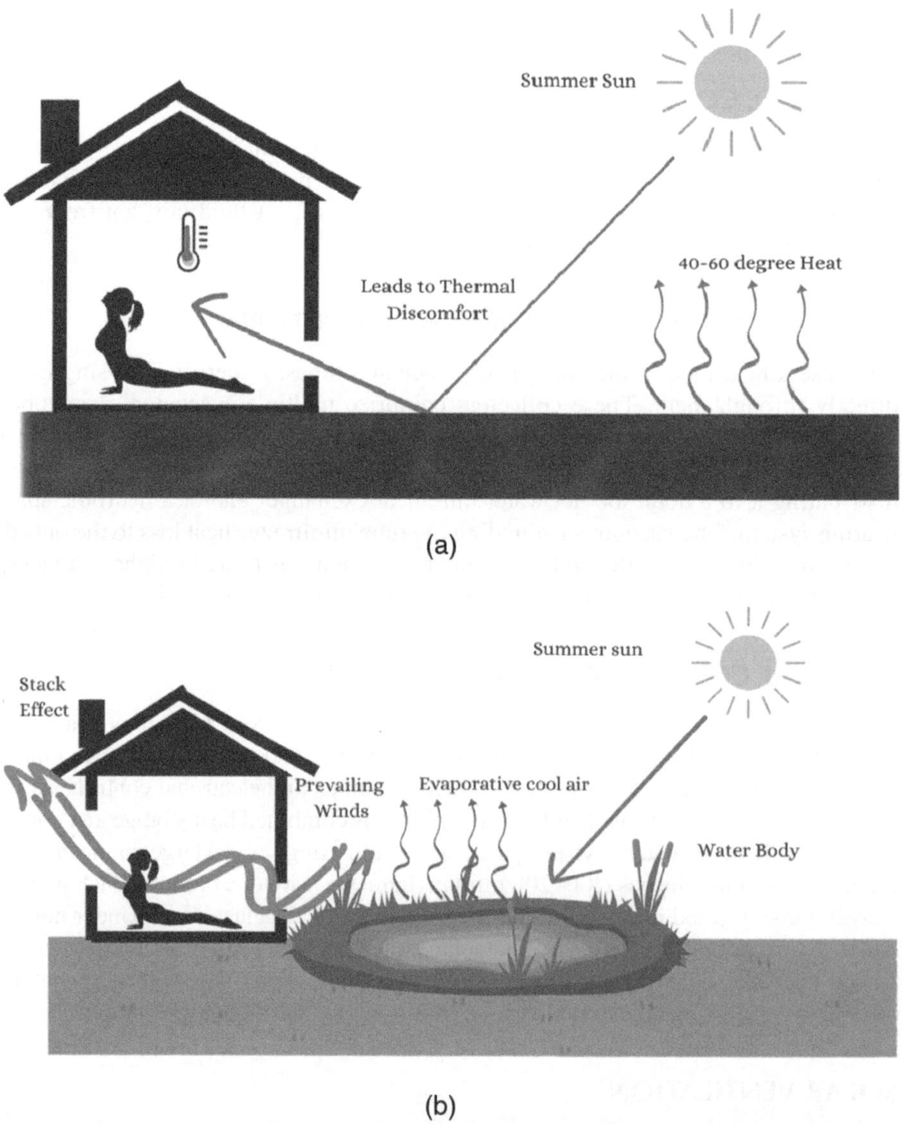

FIGURE 6.4 (a, b) Natural ventilation.

lift upwards and escape outside through openings near the roof. As the hot air parcels go upwards, cool air parcels come inside from the openings (at lower height than the openings for hot air) or windows and to fill the empty space to regulate temperature and maintain thermal comfort (Figure 6.4).

SOLAR CHIMNEY

Solar chimneys are tall, wide structures built facing the sun, made up of dark colour absorber, usually black painted surface to prove air channel (space between black surface and glass cover). During the daytime, solar radiation assisted heating that facilitated the natural stack ventilation from the top exterior vent, leads to suction of cool air from another vent at the opposite of the building, resulting in the cooling. In the night time, the top exterior vent of the chimney closed to facilitate the forced back of warm air from the chimney to the building. They can be utilized in many operations such as passive ventilation, power generation, and drying up the food products (Figure 6.5).

WIND TOWER

Wind towers are preferably used in wind areas and dry climates and have been used since ancient Egyptian times, also known as wind catcher. They are simply designed to catch and prevailing wind during the daytime through various openings at the top, leading to natural stack effects where warm air goes upwards and escapes from top opening and cool air replaces it from the lower opening in the building. In the

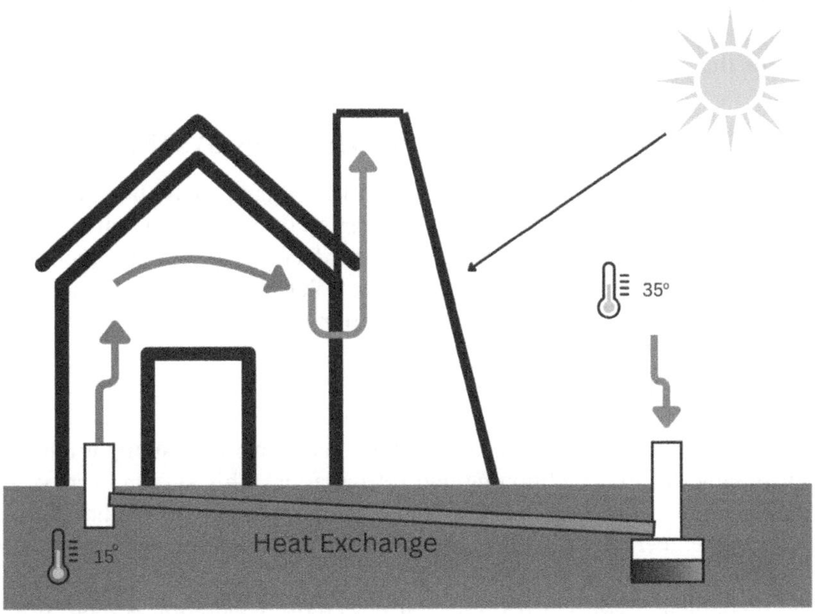

FIGURE 6.5 Solar chimney.

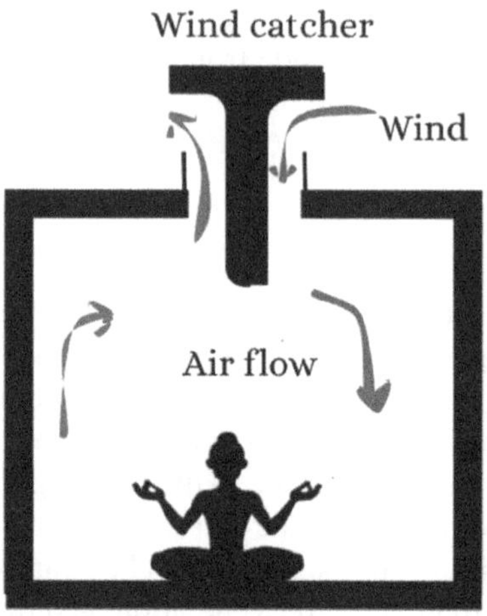

FIGURE 6.6 Wind tower.

nighttime, the cool air from the outside continuously exchanged with the indoor warm air of the building, leads to further cooling to the building (Figure 6.6).

SOLAR ROOF VENTILATION

This technique involves using solar-powered fans to extract hot air from the roof space and draw in fresh air. Solar panels on the roof power the fans, which remove stale, warm air from the attic or roof cavity and replace it with cooler outdoor air. This helps to reduce heat build-up in the roof and can also lower the overall temperature inside the building.

INTEGRATED SOLAR-POWERED VENTILATION SYSTEMS

These systems combine various components like solar panels, fans, and ducts to create a complete ventilation solution. They can be designed to automatically adjust airflow based on temperature, humidity, or other factors, optimizing ventilation performance. Examples include systems that use solar-powered fans in attics or roof cavities, or systems that preheat air for ventilation using solar radiation.

SOLAR AIR CONDITIONING

The solar cooling method consists of a system that transforms sunlight into cooling energy suitable for air conditioning and refrigeration. This system harnesses solar energy and applies it in a thermally driven cooling process. This method lowers and regulates the temperature for various applications, such as air conditioning in buildings or producing chilled water.

The fundamental concept behind solar-powered refrigeration is quite similar to traditional refrigeration. Typically, a refrigerator generates cooling by evaporating a liquid refrigerant that absorbs heat and lowers the surrounding temperature. The

TABLE 6.1
Solar cooling cycles and their feautures

Feature/cycle	Absorption	Adsorption	Desiccant	Ejector
Working Pair	LiBr–water, NH_3–H_2O	Silica gel–water, Activated carbon–methanol	Solid/liquid desiccants	Refrigerant vapour
Input Temp (°C)	80–120	50–90	50–80	90–150
COP	0.6–1.2	0.3–0.6	0.5–0.7	0.2–0.5
Maturity	Commercial	Commercial/ prototype	Commercial	Mostly prototype
Best Use Case	Large AC/ industrial	Low-temp waste heat, rural cooling	Humid climates, air dehumidification	High-temp solar or waste heat
Maintenance	Moderate (pumps, crystallization control)	Low	Low	Very low
Size	Large	Moderate to large	Compact to large	Compact

primary distinction between solar cooling and standard cooling is that the former utilizes solar thermal energy instead of electric power. This concept is practical since solar energy is generally plentiful during the summer months when the demand for air conditioning is highest. Before exploring the various cooling methods, it's crucial to clarify the difference between solar cooling and solar air-conditioning. Solar cooling lowers the temperature of a room, building, or living space by extracting heat. Solar air-conditioning, on the other hand, modifies a room or space's temperature and humidity levels to enhance occupant comfort. There are four principal types of solar-powered cooling techniques (Table 6.1)

ABSORPTION COOLING SYSTEMS

This technology is considered the oldest known cooling technology, but there is a difference in functioning, the refrigerator uses the conventional vapour compression system and relies on a mechanical compressor while a thermal compressor is employed in the absorption cooling process. The absorber in a thermal system holds a concentrated solution of coolant, whereas the generator contains a heat exchanger with a solution. In absorption chillers, the most commonly employed refrigerant-absorbent combinations are water/lithium bromide (H_2O/LiBr) and ammonia/water (NH_3/H_2O). In the water/lithium bromide pair, water acts as the refrigerant (coolant) and lithium bromide functions as the absorbent, whereas in the ammonia/water pair, ammonia is the refrigerant and water is the absorbent.

ADSORPTION COOLING SYSTEMS

The Adsorption cooling cycle is quite similar to the absorption cooling cycle, where liquid solution absorbs the refrigerant in absorption system, as in adsorption, the highly porous substance engages in the absorption of the refrigerant on its internal surface. The most commonly employed refrigerant-adsorbent combinations are activated carbon-methanol, carbon-ammonia, silica gel-water, activated carbon-ethanol, and zeolite-water. In the activated carbon-methanol pair, activated carbon acts as the solid sorbent and methanol acts as the refrigerant that adsorbs on the activated carbon.

DESICCANT COOLING SYSTEMS

In desiccant systems, desiccant materials allow for the removal of air moisture content through the sorption process as in these materials. All desiccant materials have the ability to attract moisture with different capacities. For humid climates, the desiccant cooling system proved to be more efficient in achieving thermal cooling purposes. Additionally, the low temperature gain from solar energy can also be utilized by renewable energy, waste heat, and cogeneration to drive.

EJECTOR COOLING SYSTEMS

The solar ejector-based cooling system comprises three interconnected phases: solar phase, the ejector power phase, and the ejector cooling phase. The solar phase comprises a pump, solar collectors, and a heat exchanger (generator) to transfer heat to the ejector phase. In the solar phase, solar collectors heat the working fluid, typically water, to about 95°C. Next, the heated water flows out of the solar collectors to circulate through the generator. The generator transfers heat to the working fluid of the ejector power cycle without mixing the two distinct working fluid flows, resulting in a high-pressure and high-temperature vapour.

THERMAL ENERGY STORAGE SYSTEMS

TES is one of the most practiced technologies to store energy in the form of heat to eliminate the gap between the energy supply and demand. As shown in Figure 6.1, there are three main thermal energy storage technologies; sensible heat storage through a temperature change (sensible heat) of a material, latent heat storage through phase change (latent heat) of a material and thermo-chemical heat (chemical energy) by thermally inducing changes in materials' chemical states. As compared in Table 6.1, the choice of TES method depends on a variety of factors such as the storage capacity, cost, temperature range, duration requirement as well as the specific application.

SENSIBLE HEAT STORAGE

Solid sensible heat storage is an attractive option for thermal energy storage regarding the investment and maintenance costs. Sensible heat storage stores the thermal energy by varying the temperature of storage materials, without undergoing any form of phase change within the working temperature range. Sensible heat storage could be done solid materials like rock, concrete, brick, sand; or liquid materials like water, molten salt, and thermal oil.

LATENT HEAT STORAGE

Latent heat storage utilizing PCMs is an alternative TES technique compared to the sensible heat storage. PCMs are substances which can absorb or release large amounts of energy, i.e., so-called latent heat, when they experience phase transitions among solid, liquid and gas states. Although the highest latent heat of phase change is the liquid-gas phase change, it is hard to utilize this due to the enormous volume change associated with material evaporation.

THERMO-CHEMICAL HEAT STORAGE

Different from sensible heat storage systems and latent heat storage systems, a thermo-chemical heat storage system uses reversible endothermic/exothermic reactions to absorb, store, and release energy between solar heat and chemical reactants. These reactions can be classified according to their reaction family, namely carbonate system, hydroxide system, redox system, metallic hydride system, ammonia system, and perovskite oxide system.

OFF-GRID AND GRID CONNECTED PV SYSTEM

Instead of utilizing the battery backup the system utilized the grid power for periods when solar is unavailable. In most of the cases the residential customer utilizes the rooftop solar PV system with the grid connection (Prajapati and Fernandez, 2019). Also, the grid power mostly generated with conventional fuels is nowadays installing renewable power for eco-friendly purposes (Prajapati and Fernandez, 2020, 2021a). During the solar power production period the power is consumed from the solar PV and during night times the power can be purchased from the grid. In this way the annual bill saving can be done by reducing grid power purchase (Prajapati and Fernandez, 2021b).

REFERENCES

Bansal, N.K., Hauser, G. and Minke, G. 1994. *Passive building design: A handbook of natural climatic control* (pp. 35–46). Elsevier science BV.

González-Torres, M., Pérez-Lombard, L., Coronel, J.F., Maestre, I.R. and Yan, D. 2022. A review on buildings energy information: Trends, end-uses, fuels and drivers. *Energy Reports* 8, pp. 626–637.

IEA. 2022. The future of cooling: Opportunities for energy-efficient air conditioning. https://www.iea.org/reports/the-future-of-cooling

Jain, K. and Garg, A. 2017. *Traditional architecture: House form of the Islamic community in Delhi*. Cambridge Scholars Publishing.

Jain, K. and Gautam, A. 2018. Traditional wisdom and sustainable design: Vāstu principles in Indian architecture. *Journal of Building and Environment* 134, pp. 232–243.

Kong, M., Dong, B., Zhang, R. and O'Neill, Z. 2022. HVAC energy savings, thermal comfort and air quality for occupant-centric control through a side-by-side experimental study. *Applied Energy* 306, p. 117987.

Niaki, S.O.D., Pourfallah, M. and Ghadi, A.Z. 2023. Feasibility and investigation of residential HVAC system with combined ground source heat pump and solar thermal collectors in different climates of Iran. *International Journal of Thermofluids* 20, p. 100427.

Prajapati, S. and Fernandez, E., 2019, August. Rooftop solar PV system for commercial office buildings for EV charging load. In *2019 IEEE International Conference on Smart Instrumentation, Measurement and Application (ICSIMA)* (pp. 1–5). IEEE.

Prajapati, S. and Fernandez, E., 2020, October. Performance evaluation of membership function on fuzzy logic model for solar PV array. In 2020 IEEE *International Conference on Computing, Power and Communication Technologies (GUCON)* (pp. 609–613). IEEE.

Prajapati, S. and Fernandez, E., 2021a. Capacity credit estimation for solar PV installations in conventional generation: iImpacts with and without battery storage. Energy Sources, Part A: Recovery, Utilization, and Environmental Effects 43(22), pp. 2947–2959.

Prajapati, S. and Fernandez, E., 2021b. Standalone solar PV system for grey water recycling along with electric load for domestic application. International Journal of Sustainable Engineering 14(5), pp. 933–940.

Pérez-Lombard, L., Ortiz, J., Maestre, I.R. and Coronel, J.F. 2012. Constructing HVAC energy efficiency indicators. *Energy and Buildings* 47, pp. 619–629.

Rathore, P.K.S. and Shukla, S.K. 2019. Thermal performance enhancement of phase change materials based thermal energy storage systems: A review. *Renewable and Sustainable Energy Reviews* 113, p. 109232. https://doi.org/10.2478/ama-2023-0009

Shukla, D.N. 2003. *Vastu Shastra: Ancient Indian Science of Architecture*. Roli Books.

U.S. Department of Energy. 2022. Trombe walls. Office of Energy Efficiency & Renewable Energy. https://www.energy.gov/energysaver/design/energy-efficient-home-design/passive-solar-home-design/trombe-walls

U.S. Energy Information Administration (USEIFA). 2021. Annual energy outlook 2021. https://www.eia.gov/outlooks/aeo/

Ürge-Vorsatz, D., Cabeza, L.F., Serrano, S., Barreneche, C. and Petrichenko, K. 2015. Heating and cooling energy trends and drivers in buildings. *Renewable and Sustainable Energy Reviews* 41, pp. 85–98.

7 Solar-powered water recycling

WATER SUPPLY AND WASTEWATER TREATMENT

The water supply comprises numerous interconnected elements, such as pumps, pipes, and reservoirs. Operating a standard water supply system typically demands considerable amounts of energy (Bolognesi et al., 2014). Nonetheless, energy is utilized throughout each stage of water treatment – from abstraction, conveyance, treatment, to distribution (Lam et al., 2017). The process starts at a water source, where water is extracted and transmitted, either directly to the end user or to a treatment facility, before being dispersed to consumers. Water used by end users then moves through a wastewater collection system to a treatment plant and is usually released back into the environment, not always returning to the original extraction site. Energy is essential throughout the entire water usage process. Water supply systems' energy efficiency is a pressing issue for sustainable development today. Typically, 70% of the energy used in these systems is spent on pumping, with the remaining 30% being used for purification and auxiliary supply purposes (SALGA, 2014). Water and energy conservation have become major concerns for utility companies worldwide in recent years (Mamade et al., 2014). The global need to conserve energy in water supply systems is particularly pressing, primarily due to a plan to decrease the carbon footprint, mitigate water loss from leaking pipes and boost revenue. Using energy more efficiently across the nation, and globally, leads to lower energy expenses for water companies. Increased interest in the study of an efficient energy-saving technique for WSSs can be attributed to these and other reasons.

Conventional desalination of fresh water requires energy that is typically sourced from fossil fuels (Elimelech and Phillip, 2011). Burning fossil fuels to produce energy for desalination and these processes have significant detrimental effects on the environment due to the release of greenhouse gases (GHG). By 2050, its forecast is that worldwide emissions from desalination processes will total 0.4 billion tons of CO2 equivalents annually (IEA, 2022) (Table 7.1).

WASTEWATER: TYPES AND CHARACTERISTICS

Domestic or municipal wastewater, often called sewage, encompasses the entire volume of water generated from residential areas, such as homes, flats, and communities, as well as commercial venues like stores, eateries, and inns. It is mainly made up of organic material with relatively small quantities of heavy metals and chemicals. Wastewater can be categorized into two types based on its origin: (a) grey water and (b) black water.

DOI: 10.1201/9781003652229-7

TABLE 7.1
Required energy for producing 1 cubic metre of drinking water from different water sources

Water source	Energy (kWh/m^3)
Surface water (lake or river)	0.37
Groundwater	0.48
Wastewater treatment	0.62–0.87
Wastewater reuse	1.0–2.5
Seawater	2.58–8.5

Wastewater generated from showers, sinks, and laundry is classified as grey water, which may also encompass water from dishwashing in the kitchen. As the grey water, lacking the wastewater from toilets, usually has a lower chance of harbouring pathogens, although pathogens may still be present due to sources such as baby diapers; pathogen levels are relatively less as comparison of pathogen level in lavatories' wastewater. Grey water contains residues from soap, shampoo, dead skin and hair, fats, oils, and greases, generated during showering, cloth washing, and cooking activities, although some references may exclude it from grey water. Grey water generated from sources other than toilets is generally less contaminated with organic pollutants and nitrogen compounds.

Wastewater from toilets and bathrooms is classified as black water, which include faeces, urine, water, and toilet paper often flushed in the toilet seats. It usually contains a greater amount of pathogens compared to grey water. Black water can also be categorized into brown water, which refers to wastewater from toiletries that excludes the urine, while yellow water only includes urine and water, possibly some flush water. However, there are countries like Sweden and other EU nations exploring the methods and policies to collect brown and yellow water separately. Domestic wastewater contains roughly half of the organic material load found in black water, which also comprises substantial amounts of nutrients such as N and P and can be collected with minimal water usage (about one liter per flush) through toilets that utilize vacuum systems. The effluent also contains the majority of pathogens, hormones, and pharmaceutical residues, and the quantity of black water produced varies based on the type of toilet and the amount of water used for flushing.

The characteristics of industrial wastewater or effluent differ from domestic wastewater or municipal sewage that vary in composition and characteristics with the different industry and the various processes it uses. The substance may include substantial quantities of organic waste, dyes, heavy metals, volatile substances, and synthetic chemicals, which could be hazardous and toxic in character. It can be classified depending on the type of industry where they are generated like tannery wastewater, distillery wastewater, textile wastewater, pulp wastewater and petrochemical wastewater. Various sectors have unique guidelines for the permissible levels of contaminants in their wastewater discharge.

Industrial wastewater's composition and characteristics can vary significantly throughout the day, and its discharge rate is not consistently or uniformly managed;

hence, it may necessitate an equalization process prior to treatment. The wastewater is held in a mixing tank for a specified period during the equalization phase, facilitating a more consistent flow of wastewater. In cases where the wastewater has a predominantly acidic or highly alkaline composition, a neutralization process must take place within a neutralization tank. Conducting the neutralization within the equalization tank is also feasible if the conditions are suitable.

Implementing proportioning as a unit process may be necessary when treating industrial wastewater alongside municipal wastewater, or when both types of wastewater are released into a watercourse. Regulating the discharge of industrial wastewater into a receiving sewer or stream requires maintaining a specified ratio to the flow and permissible limits of pollutants of municipal wastewater. This process offers two benefits: safeguarding treatment equipment from shock loads and improving the hygienic quality of treated wastewater.

COLLECTION AND TRANSPORT OF DOMESTIC WASTEWATER

Individual residential areas

Grey water from showers, cloth washing. kitchens (if feasible, separate collection can be done or simply passed through grease traps to remove oil and fats from the grey water) can be collected through drainage pipes and stored in storage tanks in individual houses, including those in industrial complexes. In residential houses (also in industry's residential areas), blackwater collection systems are provided with separate drainage pipes that collect wastewater from toiletries and directed to septic tanks or soak pits (installed inside the house), or sent directly to the municipality's sewerage system.

Shared residential areas (apartments, buildings, and hostels)

In apartments, flats, and hostels, grey water is collected through a common drainage system. Kitchen grey water, which includes a considerable amount of oil and grease, can impede a filtration system's performance and cause clogging. To mitigate this, kitchen grey water can be collected through a separate piping system. Grease can then be extracted from this water using a grease trap before it is either stored in a storage tank for treatment or discharged into a grey water sewerage system of the municipality.

In flats, buildings, and hostels, blackwater collection systems are provided with separate drainage pipes that collect wastewater from toiletries and directed to septic tanks or soak pits (installed inside the house), or sent directly to the municipality's sewerage system.

Commercial areas

In commercial areas, grey water is generated from excessive floor cleaning, hand basins, and food courts, but lacks the wastewater from showering and cloth washing facilities in comparison to residential areas, whereas kitchen wastewater is present in varying degrees. The commercial areas are further subcategorized into

(a) commercial areas with limited kitchen water, such as offices, institutions, and shopping complexes, etc.; (b) commercial areas with more kitchen water, including food shops, restaurants, and hotels, etc.

In commercial settings such as offices, institutions, and shopping centres, grey water from hand washing and food courts, a day-time canteen operation generates relatively small quantities of grey water, which comprises oil, grease, and detergents that are collected separately. In commercial settings such as food shops, restaurants, and hotels, grey water is primarily produced from hand basins and food courts. Grey water from hand basins can either be directed to flush toilets or stored separately for the treatment. The food courts and kitchen in commercial areas that operate during the daytime as compared to night-time (whereas residential areas show kitchen activities in night-time), led to the more grey water in day that was also stored after passing through the grease traps and treated separately.

PUBLIC PLACES (BUS STOPS, RAILWAY STATIONS, AND AIRPORTS)

Public places, such as bus stops, railway stations, and airports, produce grey water from floor cleaning, hand basins, and food courts. Grey water from hand basins is either directed to flush toilets or stored separately or released into separate sewerage

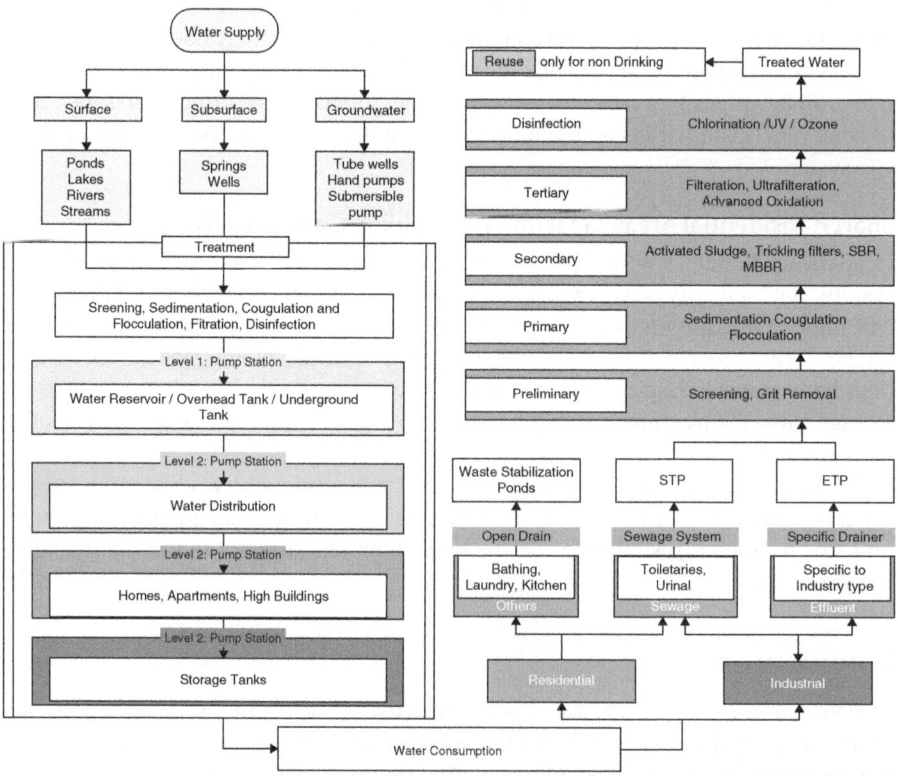

FIGURE 7.1 Water supply and wastewater treatment network.

systems for the treatment. The canteen or kitchen, which operates mainly around the clock, generates a significant amount of grey water, which is then collected and treated (Figure 7.1).

SEWAGE TREATMENT TECHNOLOGIES AND ENERGY REQUIREMENT

Many methods and technologies for sewage treatment have been developed to meet a variety of needs, from adhering to government standards having specific permissible limits for specific parameters to facilitating water recycling and reuse. The energy required for sewage treatment fluctuates significantly depending on the method or technology and several other factors related to specific methods or technologies. It is estimated that electricity consumption for the wastewater treatment sector accounts for up to 14% of the total energy consumption (Pan et al., 2011). The variability in estimated energy consumption is largely attributed to disparities in system boundaries, treatment technologies, estimation techniques, and the scale of the operations.

The type and degree of treatment needed, led to variation in electricity consumption in treatment techniques. In the Class I cities, the most implemented treatment technologies are activated sludge process (ASP) with 59.5%, Up-flow anaerobic sludge blanket (UASB) with 26% of total installed capacity. Along with these technologies, waste stabilization ponds (WSPs) are also implemented in 28% of the treatment plants, which accounts for 5.6% of their combined capacity (Central Pollution Control Board [CPCB], 2013; Kumar and Tortajada, 2020). In Class II towns, the implemented treatment technologies, WSPs with 71.9%

(72.4% of STPs) and UASB, with 10.6% (10.3% of STPs) of total installed capacity (CPCB, 2013; Kumar and Tortajada, 2020). ASP technologies are most suited and appropriate for large cities as they need less space than UASB and WSP technologies (Table 7.2).

SOLAR-POWERED WATER AND WASTEWATER TREATMENT SYSTEMS

Pumps powered by electricity or diesel are commonly employed in water pumping systems for applications like agriculture and both rural and urban water distribution. The primary issues including grid unavailability and erratic supply, as well as high costs associated with diesel pumping, necessitate a focus on pumping systems powered by renewable energy sources. Recent years have seen renewed interest worldwide in water pumping systems powered by renewable energy sources due to environmental and sustainable development concerns (Chandel et al., 2017; Wu et al., 2020). Integrating solar panels into systems for pumping water and irrigation helps to manage water resources more sustainably and efficiently (Figure 7.2).

SOLAR-POWERED PUMPS

Solar water pumps are systems that pump groundwater or surface water for irrigation, urban and rural water supply, livestock, and other uses, and they provide the power required to accomplish this through photovoltaic panels. A typical solar water pump

TABLE 7.2
Cost comparison of various technologies used in sewage treatment plants

Technology	Average capital cost (secondary treatment), million Rs/MLD	Average capital cost (tertiary treatment), Lac Rs/MLD	Total capital cost (secondary+ tertiary), million Rs/MLD	Total area (m^2) per MLD secondary+ tertiary treatment	Average total daily power requirement kWh/ day/MLD	Total annual O&M costs, Lac Rs/ MLD. Up to secondary treatment
Activated sludge process	6.80	40	10.8	1,000	185.7	353.02
Moving bed biological reactor	6.80	40	10.8	550	223.7	372.11
Sequential batch reactor	7.5	40	11.5	550	153.7	288.15
Upflow anaerobic sludge blanket + extended aeration	6.8	40	10.8	1,100	125.7	290.72
Membrane Bioreactor	30		30	450	302.5	–
Waste Stabilization ponds	2.3	40	6.3	6,100	5.7	116.09

Source: CPCB (2013b) and Kumar and Tortajada (2020).

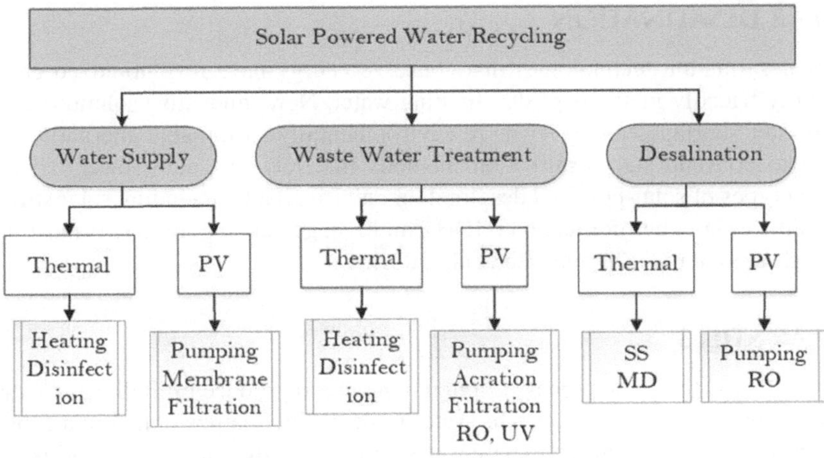

FIGURE 7.2 Solar-powered water recycling.

consists of a solar PV array, a direct current to direct current (DC–DC) converter or a direct current to alternating current (DC–AC) inverter, a motor-pump set, a water storage tank, control circuits, a suitable mounting structure, a manual or automatic tracking system, interconnect cables, protection devices, and a piping arrangement. There are three types of solar-powered water pumping systems:

a. In a solar water pumping system with a direct-coupled DC, the PV array's current and voltage are directly equivalent to the armature's current and the terminal voltage. The output power of a PV system and its pump speed change in response to the level of irradiance and the temperature of the module. When a maximum power point tracking system is integrated into the solar water pumping system, the output of the PV cells is monitored and an optimal load is applied, resulting in the maximization of power under all environmental conditions.
b. In a solar water pumping system with a direct-coupled AC, an inverter needs to be installed between the PV array and the motor. If the system generates excess energy, the surplus energy can be stored in batteries depending on the setup, or a reservoir can be installed at the pump outlet to hold pumped water. When solar radiation is unavailable, the water stored in the reservoir is utilized (Trace, 2021).
c. A solar-powered water pumping system equipped with battery backup. The incorporation of battery storage in the solar water pumping system enables the use of solar-generated power when it is required, however, this enhancement increases the overall cost of the system (Chandel et al., 2017). The photovoltaic modules are connected in series and parallel configurations to power the motor-pump unit, which is a component of solar-powered water pumping systems, and generates the necessary pressures and water flow for consumers. Solar water pumping systems, which can be employed for diverse applications, possess numerous benefits, including operational security, durability, and environmental awareness (Allouhi et al., 2019; Trace, 2021).

SOLAR DESALINATION

Solar desalination technologies that conserve energy have revolutionized environmentally friendly methods of desalinating water. New innovative solutions address freshwater shortages and offer more environmentally sustainable alternatives compared to conventional desalination methods that rely on solar power. There are several types of solar-powered desalination units available, including solar still (SS), humidification-dehumidification (HDH), multi-stage flashing, reverse osmosis (RO), and hybrid systems (Zahmatkesh et al., 2022).

SOLAR STILLS

Solar stills use the same processes found in nature to produce rainfall, which include evaporation and condensation. The water to be purified is positioned in a container within this apparatus. Solar radiation passes through the glass cover and is subsequently absorbed by the underside, which has a black coating. The absorbed radiation is immediately converted into heat. Heat from the surroundings is absorbed by seawater, and a partial evaporation of it occurs; the evaporated seawater is then condensed into distilled water on the inside of the cover. Subsequently, the drops of the distilled water start sliding downwards due to gravity and are collected at the base of the inclined transparent cover. All contaminants remain in the basin as the evaporated water leaves it (Figure 7.3).

HUMIDIFICATION-DEHUMIDIFICATION (HDH)

The humidification-dehumidification (HDH) process is analogous to the natural hydrological process initiated by the sun warming the ocean. The water in the ocean then evaporates through natural convection to form a cloud. Rainfall occurs when water is condensed from clouds. In the HDH process, a natural hydrological cycle is imitated within a confined space. Here, water evaporation occurs in the

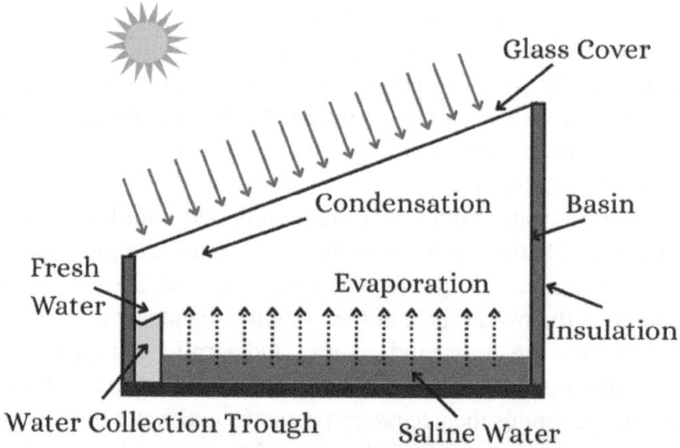

FIGURE 7.3 Solar stills.

humidification chamber, where hot water combines with dry air, and the resulting vapor is then directed to the dehumidification chamber, where a condenser is utilized to produce fresh water. The productivity of an HDH system primarily relies on glass, air, and water temperature. Productivity within the system is inversely related to air velocity.

MULTI-STAGE FLASHING

Partial evaporation or flash evaporation is the partial vapourization that occurs when a saturated liquid stream experiences a pressure drop by passing through a throttling valve or another pressure-reducing device. The flashing led to the transformation into a two-phase mixture of water and steam. Water's higher density causes it to flow downwards, while steam rises upwards. The Multi-Stage Flash (MSF) technology accounts for 21% of the overall installation capacity for water desalination. In Gulf countries with significant reserves of fossil fuels, the largest desalination plants are also based on multi-stage flash (MSF) technology. In this process, the water is initially heated and then flashed into the chambers in successive stages at varying saturation pressures (Ali et al., 2011). In solar multiple-effect distillation, the solar collectors are connected to a conventional multiple-effect distillation unit to attain high temperatures between 90°C and 120°C (Figure 7.4).

MEMBRANE DISTILLATION

Currently, several water purification methods use solar-powered membrane-based technologies to purify water through various processes, including membrane distillation (MD), and hybrid membrane systems. Porous hydrophobic and dense hydrophilic membranes are the most commonly used membrane materials for membrane-based water purification technologies. These membranes can be manufactured from both organic and inorganic materials. Organic membranes are typically fabricated from polymeric materials including polypropylene (PP), polytetrafluoroethylene (PTFE), polyvinylidene fluoride, sulfonated cross-linked polystyrene, cellulose acetate, and

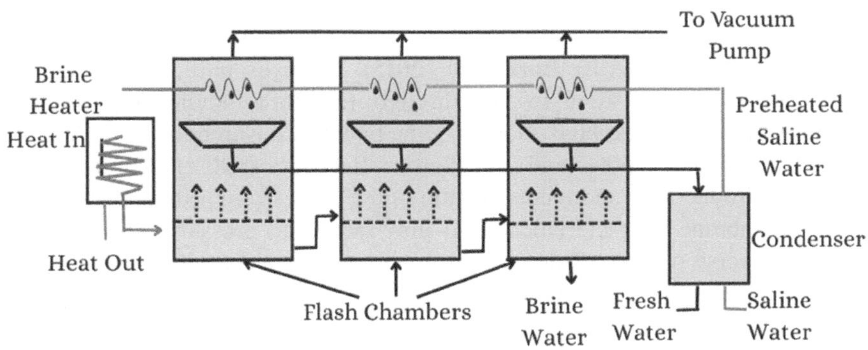

FIGURE 7.4 Multi stage flashing.

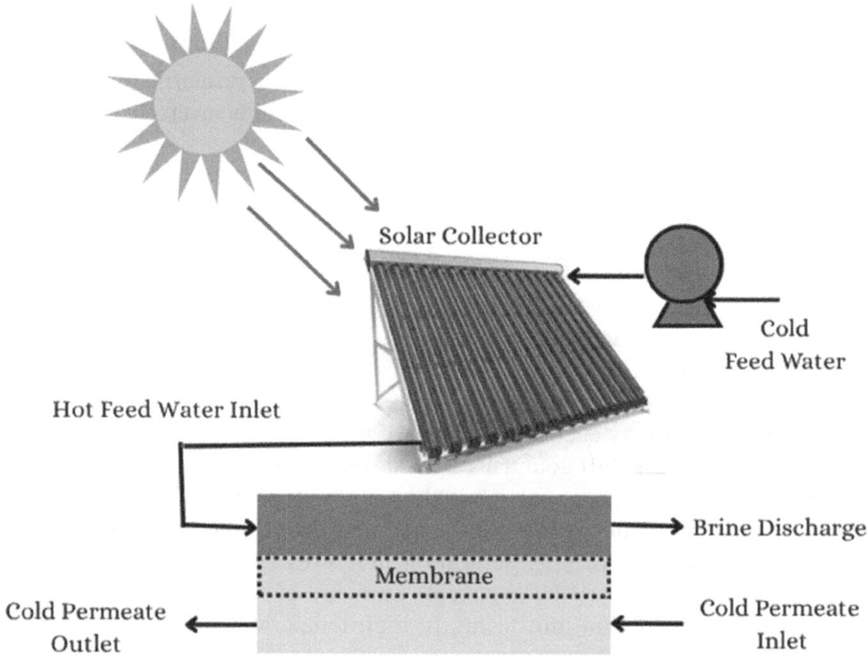

FIGURE 7.5 Membrane distillation.

aromatic polyamide. Inorganic membranes are composed of carbon, silica, zeolite, ceramic, and a variety of oxides, including alumina, titania, zirconia, and metals (Figure 7.5).

REVERSE OSMOSIS

Another efficient desalination technology is reverse osmosis, accounting for 65% of the world's total desalination installation capacity. The osmotic pressure of the feed water is decreased by passing it through a porous membrane under high-pressure conditions, specifically 8 MPa for seawater and 3 MPa for brackish water (Al-Karaghouli et al., 2010; Burn et al., 2015). The hydrophilic properties of the membrane allow only water to pass through, thereby removing impurities. The Reverse Osmosis process boasts low Specific Energy Consumption (2–5 kWhe/m^3), high recovery ratio (>50%) and high Solvent Recovery (>90%) (Ali et al., 2011). The water productivity and recovery ratio of the process are primarily determined by the membrane characteristics, feed flow rate, feed pressure, and feed water salinity. Reverse osmosis systems can be powered by electricity (generated from solar panels) or heat (produced by solar thermal collectors) or mechanical energy (harnessed by wind turbines) (Zhang et al., 2018). In PV-driven reverse osmosis systems, the photovoltaic and reverse osmosis subsystems function separately (Figure 7.6).

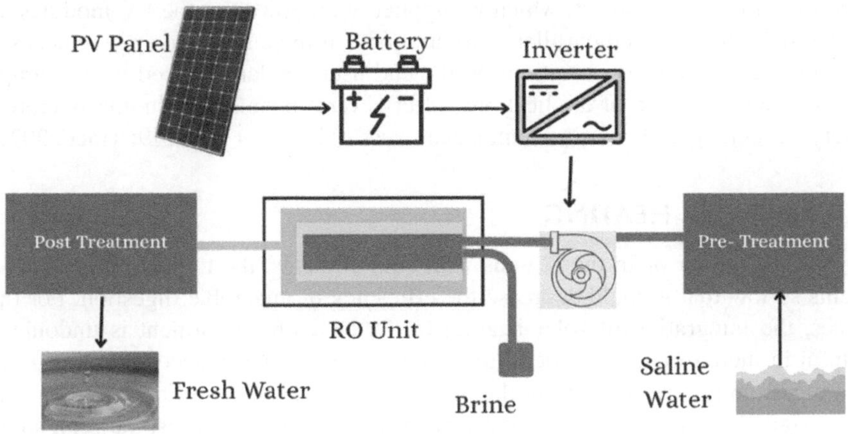

FIGURE 7.6 Reverse osmosis.

SOLAR WASTEWATER TREATMENT

Solar-powered water treatment facilities are increasingly becoming a viable solution for meeting the growing demand for clean water. These facilities reduce their reliance on conventional energy sources by tapping into solar energy, resulting in more eco-friendly water treatment processes.

SOLAR PUMPING

A typical solar water pump system comprises a solar PV array, a DC–DC converter or a DC–AC inverter, a motor-pump set, a water storage tank, control circuits, a suitable mounting structure, a manual or automatic tracking system, interconnect cables, protection devices, and a piping arrangement. There are three types of solar-powered water pumping systems.

In a direct-coupled DC solar water pumping system, the PV array's current and voltage are equivalent to the armature current and terminal voltage. The output power of the PV system and the speed of the pump are both influenced by the level of irradiance and the module's temperature. When a maximum power point tracking system is integrated into the solar water pumping system, the output of the photovoltaic cells is monitored and the optimal load resistance is applied, thereby maximizing the obtained power under all environmental conditions.

In a solar water pumping system that is direct-coupled to AC, an inverter needs to be fitted between the photovoltaic array and the motor. Unused energy produced by the system can be stored in batteries, or a reservoir can be installed at the pump outlet to store water. In the absence of solar radiation, water stored in the reservoir is utilized (Trace, 2021).

A solar-powered water pumping system with battery backup. The integration of battery storage into the Solar water pumping system enables the utilization of solar-generated power at the point of consumption, however this approach escalates the system's overall cost (Chandel et al., 2017). The solar water pumping systems comprise a motor-pump set, which is supplied with power by the PV modules connected in both series and parallel configurations, ultimately producing the necessary pressures and water flow required by the end-users. Solar-powered water pumping systems, with a range of applications, offer several benefits, including operational safety, reliability, and environmental awareness (Allouhi et al., 2019; Trace, 2021).

SOLAR WATER HEATING

During the winter or in the regions with cold climates, the temperature of sewage seems so low that it could decrease the efficiency of anaerobic digestion. For these issues, the integration of solar heating in the anaerobic treatment is undoubtedly helpful in increasing the temperature of sewage; resulting in acceleration of microbial digestion process to achieve the higher sewage treatment efficacy. According to the design of solar energy anaerobic wastewater treatment systems, they can address the energy issue in anaerobic treatment heating systems for regions with abundant solar energy. Utilizing solar pre-heated sewage in anaerobic pools can address the low-temperature and freezing issues associated with wastewater, which is brought about by significant temperature disparities in plateau regions. Yiannopoulos (2011) employed Anaerobic Biofilter reactors to elevate the sewage treatment temperature via solar heating. The data indicate that solar heating can achieve the intermediate temperature (35°C) required for anaerobic reactions. The prospects for solar heating anaerobic biological treatment appear to be favourable. Ren et al. (2012) outlines the technical approaches of solar wastewater treatment systems and suggests that advancements in anaerobic biological treatment should be pursued through the advancement and utilization of solar heating systems.

SOLAR AERATION

Various studies have suggested that using photovoltaic power generation to provide electricity for sewage biological treatment can reduce the energy usage of sewage treatment facilities. Sewage treatment plants are known for their high electric load and substantial energy consumption, operating continuously 24 hours a day with a consistent load. A photovoltaic power system can fulfill the needs of a water treatment facility in alignment with the "self-sufficiency" approach. Electricity costs represent more than 30% of production expenses, necessitating significant efforts to decrease expenses and enhance efficiency. The integration of photovoltaic power with wastewater treatment, along with adopting contract energy management, can further reduce wastewater treatment costs. There are already some practical examples of using solar energy as a power source for sewage treatment. Hudnell et al. implemented a solar-powered circulating agitation system as an alternative to traditional aeration methods in oxidation ponds, effectively reducing electricity consumption and operational costs while also deodorizing and minimizing sludge generation and greenhouse gas emissions. Han et al. employed solar power to run the oxidation ditch without a battery, with the solar power system automatically starting and stopping based on changes in light intensity.

SOLAR SLUDGE DE-WATERING AND DRYING

- Using Solar Greenhouses: A greenhouse-type solar drying device boasts a straightforward design and uncomplicated usage. Typically, materials applied to the drying device are chosen for their high transmittance in order to receive more solar radiation. To facilitate drying, a sufficiently sized space is necessary for the construction of the drying chamber, thus ensuring the sludge receives sufficient sunlight to achieve low water content.
- Utilizing a Heat Collector and Storage System: To enhance the stability of the drying system, a collector solar drying system can be employed. The solar collector is employed to transform solar energy into electricity, which is capable of heating water at a consistent temperature, thereby enabling hot water to thermally dry the sludge through floor radiation (Müller and Saucer, 2012).
- Using Solar Greenhouses with Heat Pumps: The sludge drying technology, which incorporates both solar energy and a heat pump, is able to maximize the use of the heat-pump system, thereby achieving high evaporation temperatures with high efficiency. A solar collector provides the benefits of effective heat collection at low temperatures (Tang et al., 2020). This new combination system effectively addresses the instability issue associated with solar energy. At the same time, it achieves the objectives of energy saving, emission reduction, and pollution prevention (Wang et al., 2021) (Figure 7.7).

SOLAR PHOTO-CATALYSIS

Photocatalysis is employed to accelerate the photoreaction process through the use of a catalyst. This technique serves as an efficient method of oxidation for the mineralization of water pathogens and persistent organic compounds. There are two primary

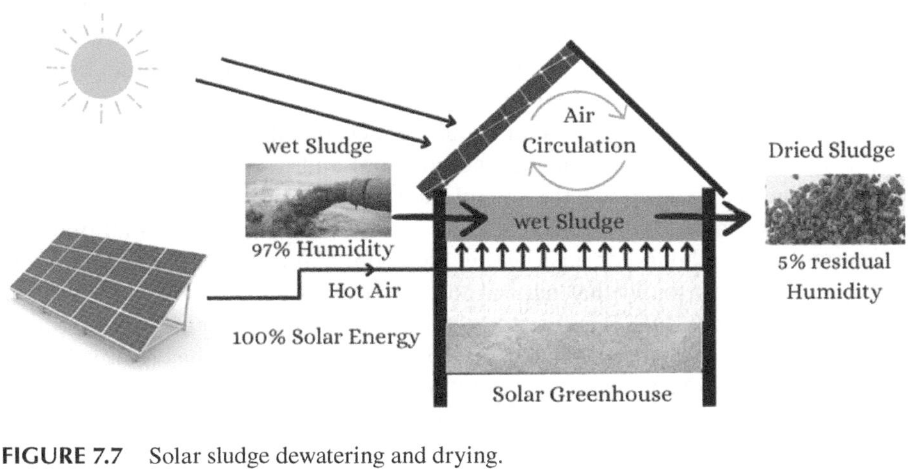

FIGURE 7.7 Solar sludge dewatering and drying.

types of photocatalytic processes: heterogeneous and homogeneous photocatalysis. Numerous detailed review articles summarize the essential features of heterogeneous and homogeneous photocatalysis (Byrne et al., 2015; Ibhadon and Fitzpatrick, 2013; Chong et al., 2010; Pelizzetti and Serpone, 2012; Malato et al., 2002). In heterogeneous photocatalysis, the oxidation of organic substances in water takes place in the presence of a photocatalyst and UV light. Various reactive oxygen species are produced that not only eliminate water pollutants but also break down microorganisms (Blake et al., 1999). Among the various photocatalysts (such as TiO2, ZnO, Fe2O3, ZnS, CdS, etc.), TiO2 has received considerable attention and is frequently utilized in water treatment applications. Conversely, homogeneous solar photocatalysis is linked to higher oxidation states (particularly those of metal ions) and the creation of hydroxyl radicals under both light and heat conditions. Consequently, this method is often employed as catalysts for metal complexes like Fe, Cu, Cr, and others. In wastewater treatment scenarios, the photo-Fenton process (also known as Fenton oxidation) is a commonly used homogeneous photocatalytic technique. In industrial applications, typical photoreactors include both concentrating and non-concentrating varieties of solar collectors (Qin et al., 2017). The specific collectors mentioned include parabolic trough concentrators, compound parabolic concentrators, Scheffler dishes, and solar towers.

SOLAR WATER DISINFECTION (SODIS)

This technology dates back to ancient times and was first employed by inhabitants of the Indian subcontinent approximately 2,000 years ago. The simplest and most cost-effective household water treatment process after boiling and filtration is available. In the last decade, SODIS has regained scientific attention and has been promoted by the WHO, especially in countries where there is a lack of access to drinkable water but solar energy is readily available. The SODIS process is affected by several factors, including UV radiation, water temperature, the properties of the bottle, water quality, and external factors (Vivar et al., 2015). To prevent the re-growth of bacteria after exposure, solar disinfected water should be consumed within 24 hours. In SODIS, the water is stored in a transparent container and left out in the sunlight to kill waterborne pathogens like bacteria, viruses, and protozoa (McGuigan et al., 2012). Bacterial inactivation results from the sequential disruption of normal cellular processes and the cessation of enzymatic synthesis. The loss of membrane potential and decrease in glucose levels leads to the permeability of the cytoplasmic membrane and loss of cultivability. Of all waterborne pathogens, bacteria are highly responsive to solar disinfection, requiring six hours for inactivation under optimal environmental conditions. Some studies have found that natural fecal coliforms and certain light-resistant forms of cultured E. coli inactivated at slower rates (Sommer et al., 1997; Sinton et al., 2002). Compared to light sensitive organisms, the letter has relatively slow inactivation rates (McGuigan et al., 1998). Bacterial organisms such as bacillus subtilis are neither waterborne pathogen nor do them form complex microorganisms, but instead form them when exposed to sunlight in the SODIS process. Improved SODIS has been shown to be effective for disinfecting potable water using solar photo-catalytic (TiO_2) technology. TiO_2 photo-catalysis has been successfully utilized to inactivate microorganisms that are particularly resistant to UV-A radiation.

REFERENCES

Ali, M.T., Fath, H.E.S. and Armstrong, P.R. 2011. A comprehensive techno-economical review of indirect solar desalination. *Renewable and Sustainable Energy Reviews* 15(8), pp. 4187–4199.

Al-Karaghouli, A., Renne, D. and Kazmerski, L.L. 2010. Solar and wind opportunities for water desalination in the Arab regions. *Renewable and Sustainable Energy Reviews* 13(9), pp. 2397–2407.

Allouhi, A., Saadani, R., Kousksou, T., Jamil, A. and Zeraouli, Y. 2019. Technical and economic performance optimization of solar water pumping systems. *Energy Conversion and Management* 166, pp. 402–414.

Blake, D.M., Maness, P.C., Huang, Z., Wolfrum, E.J., Huang, J. and Jacoby, W.A. 1999. Application of the photocatalytic chemistry of titanium dioxide to disinfection and the killing of cancer cells. *Separation and Purification Methods* 28(1), pp. 1–50.

Bolognesi, A., Bragalli, C., Lenzi, C. and Artina, S. 2014. Energy efficiency optimization in water distribution systems. *Procedia Engineering* 70, pp. 181–190.

Burn, S., Hoang, M., Zarzo, D., Olewniak, F., Campos, E., Bolto, B. and Barron, O. 2015. Desalination techniques—A review of the opportunities for desalination in agriculture. *Desalination* 364, pp. 2–16.

Byrne, J.A., Dunlop, P.S.M., Hamilton, J.W.J., Fernándes, R. and Araujo, R. 2015. A review of heterogeneous photocatalysis for water and surface disinfection. *Molecules* 20(4), pp. 5574–5615.

Central Pollution Control Board (CPCB). 2013. Performance evaluation of sewage treatment plants. https://cpcb.nic.in/openpdffile.php?id=UmVwb3J0RmlsZXMvMjlfMTQ1ODExMDk5Ml9OZXdJdGVtXzE5NV9TVFBfUkVQT1JULnBkZg

Chandel, S.S., Naik, M.N. and Chandel, R. 2017. Review of solar photovoltaic water pumping system technology for irrigation and community drinking water supplies. *Renewable and Sustainable Energy Reviews* 49, pp. 1084–1099.

Chong, M.N., Jin, B., Chow, C.W.K. and Saint, C. 2010. Recent developments in photocatalytic water treatment technology: A review. *Water Research* 44(10), pp. 2997–3027.

Elimelech, M. and Phillip, W.A. 2011. The future of seawater desalination: Energy, technology, and the environment. *Science* 333(6043), pp. 712–717.

Ibhadon, A.O. and Fitzpatrick, P. 2013. Heterogeneous photocatalysis: Recent advances and applications. *Catalysts* 3(1), pp. 189–218.

IEA. 2022. The future of water desalination: Energy and emissions projections to 2050. https://www.iea.org/reports/net-zero-by-2050

Kumar, M.D. and Tortajada, C., 2020. Wastewater Treatment Technologies and Costs. In: Assessing Wastewater Management in India (pp. 35–42). Springer, Singapore.

Lam, K.L., Kenway, S.J. and Lant, P.A. 2017. Energy use for water provision in cities. *Journal of Cleaner Production* 143, pp. 699–709.

Malato, S., Blanco, J., Vidal, A. and Richter, C. 2002. Photocatalysis with solar energy at a pilot-plant scale: An overview. *Applied Catalysis B: Environmental* 37(1), pp. 1–15.

Mamade, A., Loureiro, D., Covas, D. and Alegre, H. 2014. Energy auditing as a tool for improving service efficiency of water supply systems. *Procedia Engineering* 89, pp. 557–564.

McGuigan, K.G., Joyce, T.M., Conroy, R.M., Gillespie, J.B. and Elmore-Meegan, M. 1998. Solar disinfection of drinking water contained in transparent plastic bottles: Characterizing the bacterial inactivation process. *Journal of Applied Microbiology* 84(6), pp. 1138–1148.

McGuigan, K.G., Conroy, R.M., Mosler, H.J., du Preez, M., Ubomba-Jaswa, E. and Fernandez-Ibanez, P. 2012. Solar water disinfection (SODIS): A review from bench-top to roof-top. *Journal of Hazardous Materials* 235–236, pp. 29–46.

Müller, J. and Sauer, A. 2012. Solar drying of sewage sludge in thin layers: A review. *Renewable and Sustainable Energy Reviews* 16(1), pp. 1–8.

Pan, T., Zhu, X.D. and Ye, Y.P., 2011. Estimate of life-cycle greenhouse gas emissions from a vertical subsurface flow constructed wetland and conventional wastewater treatment plants: A case study in China. *Ecological Engineering*, 37(2), pp. 248–254.

Pelizzetti, E. and Serpone, N. 2012. Homogeneous and heterogeneous photocatalysis: From water photolysis to applications in environmental cleanup. *Catalysis Today* 77(1–2), pp. 1–12.

Qin, D., Wang, X. and Zhang, J. 2017. Application of solar photoreactors in wastewater treatment: A review. *Renewable and Sustainable Energy Reviews* 76, pp. 1120–1134.

Ren, Z., Chen, Z., Hou, L., Wang, W., Xiong, K., Xiao, X. and Zhang, W. 2012. Design investigation of a solar energy heating system for anaerobic sewage treatment. *Energy Procedia* 14, pp. 255–259.

SALGA. 2014. Guideline on energy efficiency and renewable energy in municipal water and wastewater infrastructure, SALGA energy guidelines series. South Africa, 01–23, 2014.

Sinton, L.W., Hall, C.H., Lynch, P.A. and Davies-Colley, R.J. (2002). Sunlight inactivation of fecal indicator bacteria and bacteriophages from waste stabilization pond effluent in fresh and saline waters. *Applied and Environmental Microbiology* 68(3), pp. 1122–1131.

Sommer, B., Marino, A., Solarte, Y., Salas, M.L., Dierolf, C., Valiente, C., Mora, D., Rechsteiner, R. and Wegelin, M. 1997. SODIS: An emerging water treatment process. *Journal of Water Supply: Research and Technology—AQUA* 46(3), pp. 127–137.

Tang, Y., Liu, H. and Zhao, Y. 2020. Integration of solar and heat pump technologies for energy-efficient sludge drying. *Energy and Buildings*, 210, p. 109736.

Trace, T. 2021. Solar water pumps: A clean solution for agriculture and drinking water. Solar Development Reports. https://www.tracesustainability.org/solar-water-pumping

Vivar, M., Fuentes, M. and Pujol, J. 2015. Factors affecting the efficiency of solar water disinfection (SODIS) using PET bottles: A review. *Renewable and Sustainable Energy Reviews* 41, pp. 1274–1291.

Wang, F., Zhang, T. and Li, Y. 2021. High-performance hybrid solar-heat pump sludge drying system: Design and operational assessment. *Journal of Cleaner Production* 279, p. 123659.

Wu, X., Zhang, X. and Lu, Y. 2020. Renewable energy powered sustainable water pumping systems: Developments, challenges, and future prospects. *Renewable Energy* 147, pp. 1780–1797.

Yiannopoulos, G. 2011. Anaerobic biofilter reactors combined with solar heating for wastewater treatment. *Desalination* 273(1), pp. 160–165.

Zahmatkesh, S., Kassian, T.T., Amesho, M.S. and Chongqing, W. 2022. Integration of renewable energy in wastewater treatment during COVID-19 pandemic: Challenges, opportunities, and progressive research trends. *Cleaner Chemical Engineering* 3, p. 100036.

Zhang, Y., Ma, Z. and Wang, C. 2018. Renewable energy powered membrane technology: A review of sustainable approaches for water treatment. *Renewable and Sustainable Energy Reviews* 82, pp. 3833–3855.

8 Solar-powered transportation

HISTORY OF TRANSPORTATION AND ENERGY SOURCE

The evolution of transportation and adoption of new fuels or energy sources always remain parallel to each other; having a transitional journey of wood to coal to crude oil. Prior to the industrial revolution, which took place in the 18th century, energy consumption was heavily dependent on animal power and firewood resources. The majority of the activities were carried out using manual labour and livestock, with biomass, predominantly firewood, being utilized to fulfil cooking and heating related energy needs. Although there were alternative energy sources like watermills and windmills available, their overall presence was relatively limited and primarily used for particular operations, such as grinding grain into flour (Smil, 2017). Later a substantial transition took place in the field of energy resources with coal use spreading (Fouquet and Pearson, 2012), primarily for steam engines, but increasingly for power stations to support the industrial revolution in the 19th century. This transformation was driven primarily by the conversion of thermal energy into mechanical energy. The majority of the events occurred in regions close to coal mines. Coal was the primary source of energy, but a gradual transition to higher energy content fuels such as oil eventually occurred. The second major shift resulted in the invention of internal combustion engines in the late 19th century and the increasing trend of oil-powered vessels (Gales et al., 2007).

In the 20th century, transportation has become over-reliant on petroleum sources and it also becomes the primary source of energy in the global economy. This trend fuels the evolution of a comprehensive petroleum distribution network encompassing pipelines, storage tanks, oil tankers, and filling stations (Yergin, 2012). Later with the help of exploration and technological advancement, more efficient sources of fossil fuels like natural gas were exploited, and a completely new type of energy, nuclear energy through nuclear fissions, became accessible (Smil, 2010). Hydroelectric, wind, and solar alternatives began to be utilized, yet they continued to be minor energy sources (Fouquet, 2014).

The 21st century experienced significant changes in energy sources, including the gradual replacement of fossil fuels like coal and oil with more efficient alternatives, including natural gas and other renewable resources such as wind and solar energy (IEA, 2021). Developments in biotechnology highlight the increasing potential of fuels produced from biomass like biogas, methane, and energy from waste which are also set to account for a significant portion of energy sources (Demirbas, 2009; Scarlat et al., 2015). Beside the adoption of renewable energy and advancements, the transportation sector is still responsible for 23% of the total energy consumption and 15% of carbon emissions worldwide in the year 2020 (Jaramillo et al., 2022). The

DOI: 10.1201/9781003652229-8

transportation sector includes the mobility from the road, rail, air, and sea and also comprises transport of either passengers or freight. Shukla et al. (2022) estimated that 70% of direct carbon emissions emitted through road vehicles, 12% emission from aviation, 11% emission from shipping and 1% emission from rail.

The adoption of electric vehicles (EVs) is becoming a new trend to reduce dependency on fossil fuels, lowering the transportation-related carbon emissions and becoming carbon-neutral. IEA (2024a) presented the 2024 scenario, the worldwide fleet of EVs utilized approximately 180 TWh of electricity, representing an increase of nearly 60% from the year before. On a global scale, EVs accounted for nearly 0.7% of total electricity consumption in 2024. The number of EVs is projected to grow to more than three times its current size by 2030, yet electricity demand is anticipated to rise more than four times, reaching 780 TWh in the STEPS scenario. This surge is fuelled by higher consumption from electric trucks and a greater adoption of EVs in regions where annual driving distances are more substantial. Overall energy demand for road transport is expected to rise by just 5% in 2030 in the STEPS, while total road activity (in terms of vehicle kilometres travelled) increases by almost 20% during the same period, reflecting the greater energy efficiency of EVs (IEA, 2024b).

ELECTRIC VEHICLES AND CHARGING INFRASTRUCTURE

EVs have gained popularity towards a safe environment and are more energy-efficient, becoming an integral part of modern transportation among various developed technologies. The EVs are typically categorized into three groups based on the electricity source utilized for the propulsion: (a) battery-powered electric vehicles (BEVs): includes only electric propulsion systems that are powered solely by battery and they need larger batteries compared to HEV and PHEV to provide power to the vehicle; (b) hybrid electric vehicles (HEVs); includes both the

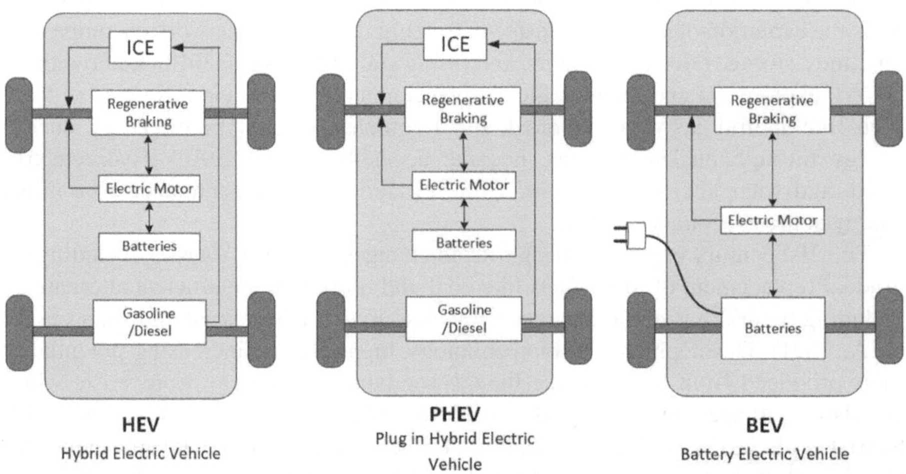

FIGURE 8.1 Types of electric vehicles.

electric propulsion system and internal combustion engine, have the choice of either running on electric or fossil fuel; led to increase fuel efficiency and lower carbon emissions of conventional internal combustion engine vehicles (ICEVs), but they have smaller batteries suitable for short duration in electric mode especially for urban use; (c) plug-in hybrid electric vehicles (PHEVs): They also have electric propulsion systems and powered by rechargeable batteries along with internal combustion engines for back up like in the HEVs but they have larger batteries to provide longer duration in electric mode and once the battery is depleted, a gasoline engine serves as a backup power source. (c) Fuel cell EVs: includes an electric propulsion system that relies on fuel cells rather than batteries or a battery in conjunction with a super capacitor (Figure 8.1).

There are several methods to recharge a battery of EVs and regulate its flow of current to the electric propulsion system. EVs employ rectifiers to transform AC into DC for recharging their battery packs. The charging methods are classified into inductive charging, conductive charging, and battery swapping.

INDUCTIVE CHARGING OR WIRELESS POWER TRANSFER

This technology utilizes a two-coil system based on electromagnetic induction. Once the installation is finished, the receiving coil is fitted to the vehicle, while the charging coil is positioned on the road surface. Due to technological developments, wireless power transfer techniques have gotten the attention in EV charging as inductive charging proved to be safe and convenient recharging of the vehicle. Although this technique doesn't require a standard connector for the charger, a standard coupling technology must be present to avail the charging facility. The biggest advantage of inductive technology is that the charger can be used while the vehicle is in motion. Despite the many advantages, inductive charging or wireless power transfer is not particularly efficient.

CONDUCTIVE CHARGING

The conductive charging method categorized into: (a) Depot charging and (b) pantograph charging. The advantages of conductive charging include its cost-effectiveness, rapid charging capability, ease of use, and high level of efficiency. Besides overnight depot and pantograph and charging systems, conductive charging has also been categorised into: On-board charging and Off-board charging based on the position of the charger either present inside the vehicle or outside the vehicle.

a. **Depot charging:** This includes facilities that provide charging in depots especially for EVs used for offices and institutions, which are often used during the day and for limited range. It is commonly employed for charging during the evening hours. Depot charging offer both fast and slow charging options through: On-board charger, where EV chargers fixed on the vehicle are small in size and limited power due to space and size constraints led to slow charging; Off-board charger, present outside the vehicle that offers fast recharging capabilities due to no size and space constraints.

b. **Pantograph charging:** This charging method is usually employed for vehicles demanding greater battery capacity and increased power, like buses and trucks. This charging technique leads to lower investment requirements for bus batteries, resulting in decreased investment costs, while increasing the costs of charging infrastructure. The charging method setup could be classified into (a) Top-down pantograph: The pantograph equipment is typically situated above the bus stop; also known as Off-board top-down pantograph. This technique often utilizes direct current to generate high power; (b) Bottom-up Pantograph: This charging method is suitable for buses that are already equipped with charging equipment. The bottom-up pantograph is also known as the onboard pantograph.

BATTERY SWAP STATIONS

Battery swap stations facilitate battery swaps via a process known as "Battery Exchange," where customers rent batteries from the battery swapping facility for a fixed duration like monthly or yearly basis. These charging stations not only prolong the charging process but also extend the lifespan of batteries. Battery swap stations could be developed with solar power to reduce the dependency on grid power and can be easily integrated into the BSS system. The advantages of these swap stations; vehicle owner doesn't need to leave the vehicle for charging for a longer time and the battery can be replaced rapidly (Figure 8.2).

EV charging stations are primarily refuelling stops that provide electric power to charge or swap batteries of EVs. The charging facility includes an interface panel that is integrated with a charging cable, charging port, and control panel. The power outlet configuration depends on a grid configuration, with parameters including voltage rating, frequency rating, and transmission standards playing a determining role. Established standards outline various charging types, encompassing

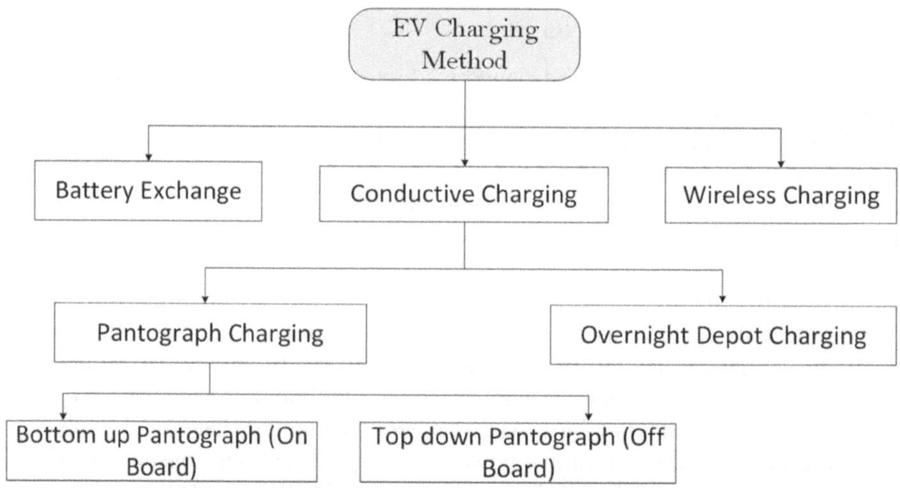

FIGURE 8.2 EV charging methods.

alternating current (AC) Level-1, alternating current (AC) Level-2, and direct current fast charging. Schroeder and Traber (2012) found that residential charging infrastructure is the most widespread method for charging EVs, with Level-1 and Level-2 being the most frequently used.

Level 1 charging cables, which are classified as L1, come with every EV. The device is universally compatible, does not necessitate installation fees, and can be plugged into any standard grounding 120-V outlet.

Level 2 charging stations are often found in public and residential areas, such as those that use Level 2 (L2) technology. To meet L2 requirements for EVs, charging stations need to employ a single-phase power supply of 240 V with a maximum 40 A current flow capacity for residential and commercial setups, and a three-phase AC power supply of 400 V with a maximum 80 A current flow capacity.

Level 3 charging is the quickest method for EV charging, classified as Level 3 (L3). The primary function of the L3 charging station is to charge EVs in both public and commercial zones. Vehicles can be rapidly recharged in busy locations like rest stops, shopping malls, and entertainment districts thanks to their quick refuelling abilities.

SOLAR-POWERED TRANSPORTATION

The fuel costs for conventional petrol or diesel-based vehicles are increasing, making EVs the future mode of transportation. Although the adoption of EVs helps in reducing the usage of fossil fuel in the future; but it also fuels the increasing electricity demand. To provide alternative energy resources, solar power panels, known as photovoltaic (PV) panels, can be used to charge EVs in an environmentally friendly manner. Harnessing office building solar potential for EV charging at work, and capitalizing on the extended parking periods, paves the way for the widespread adoption of vehicle-to-grid (V2G) technology (Figure 8.3).

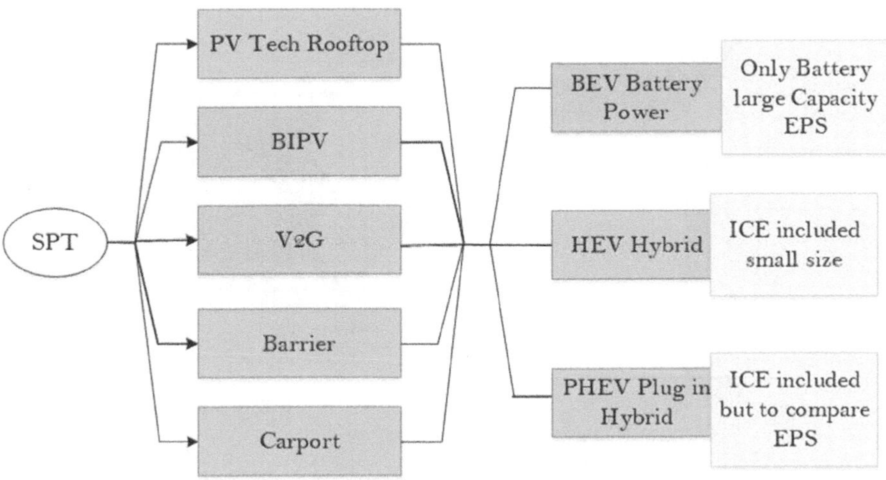

FIGURE 8.3 Solar-powered transportation.

PHOTOVOLTAIC NOISE BARRIERS

A photovoltaic noise barrier (PVNB), also known as a solar noise barrier, involves the integration of noise barrier and solar PVs for the direct conversion of sunlight into electrical power. PVNBs can be developed through retrofitting of PV modules into existing noise barriers or the integration of these modules into the design of new noise barriers. In both the cases, the noise barrier acts as a base substructure for PV modules. Top-mounted retrofit designs that offer extra space to an existing noise barrier structure are currently the most prevalent approach for the PVNBs. PVNBs provide a clever solution for tackling two issues concurrently. Noise pollution from busy roads and train tracks is reduced by these systems, which also facilitate the production of renewable energy. PV panel installations are integrated into noise barriers along highways, railways, and other noisy settings. This helps reduce carbon emissions and promotes greater reliance on renewable energy sources. Noise barriers can be upgraded by incorporating solar panels, which not only reduce noise pollution but also aid in renewable energy generation.

Two main types of PVNB designs are available – a retrofit design and a vertical bifacial design. In the retrofitting design, conventional polycrystalline PVs are mounted on existing noise barriers, whereas in the vertical bifacial design, PVs are arranged between two panes of thick toughened glass. Solar cells on both sides of the barrier are able to receive sunlight penetration. In recent times, technologies like PV glass, thin-film PV, semi-transparent PV, and luminescent solar concentrators (LSCs) have been tested in noise barrier applications to identify novel methods for integrating solar PV with existing infrastructure on a larger scale. Further new technologies related to use of different manufacturing materials are facilitating new dimensions regarding the colours and shapes (Figure 8.4).

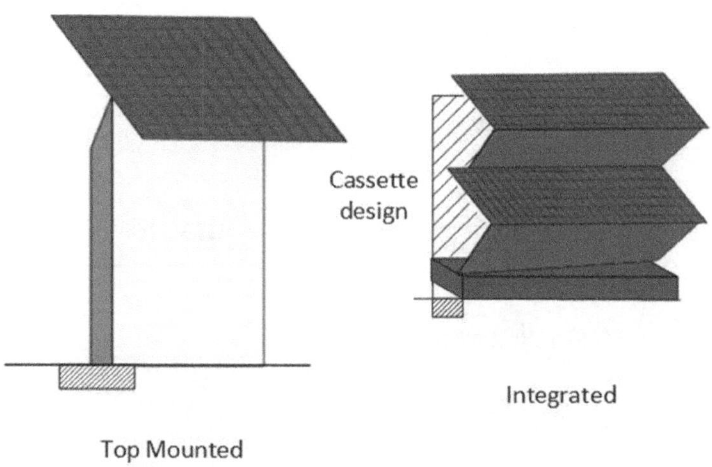

FIGURE 8.4 Photovoltaic noise barriers.

SOLAR CAR PARKING OR CARPORT

A Solar Carport is a self-sufficient dual charging station that doesn't need an external power source. It features a PV setup with solar panels and built-in batteries. This system offers protection and shade, thereby lowering the expense of electricity bill payments. Exposed car parks are transformed into power stations with the solar canopies capturing abundant free energy directly from the sun (Figure 8.5).

VEHICLE-TO-GRID TECHNOLOGY

V2G integration represents a groundbreaking concept in the fields of energy and transportation as EVs converge with the power grid. This new perspective on vehicular energy consumption allows EVs to integrate seamlessly with the power grid, moving beyond their traditional role as vehicles (Ustun et al., 2013). V2G integration involves a sophisticated framework specifically designed to enable bi-directional electricity transfer between EVs and the power grid. At its core, the system features the EV, which acts as a mobile energy storage unit (Zhang et al., 2020). It is equipped with a battery management system (BMS) responsible for overseeing and regulating both charging and discharging functions (Long et al., 2022). The Electric Vehicle Supply Equipment, connected to the EV, comprises a bidirectional charger that permits electricity to flow in both directions. Additionally, it includes a communication interface that facilitates data exchanges among the EV, the grid, and the user. The process of interfacing with the grid requires a transformer to manage voltage levels, a smart meter for accurate monitoring of energy transfers, and the local distribution network for electricity delivery (Yang et al., 2024). An aggregator platform manages the coordination of multiple EVs' combined energy resources. This platform interacts with the grid operator to improve power grid efficiency by offering services such as lowering peak energy demand and stabilizing the grid frequency. A robust communication network is essential for secure and real-time data sharing, compliant with standards such as ISO 15118 and IEEE 2030.5. The grid management system, comprising an Energy Management System (EMS) and Distribution Management

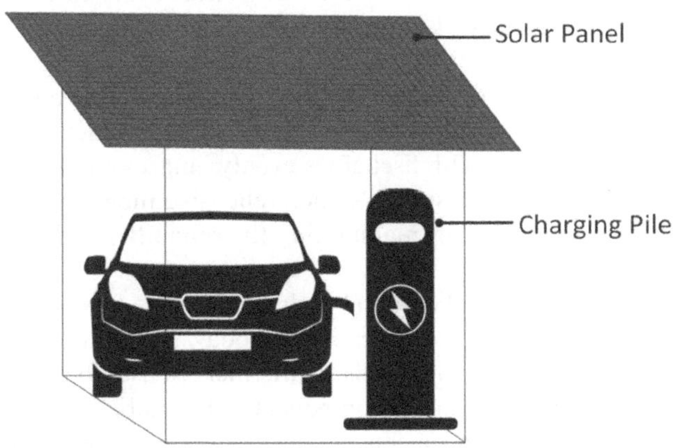

FIGURE 8.5 Solar carport.

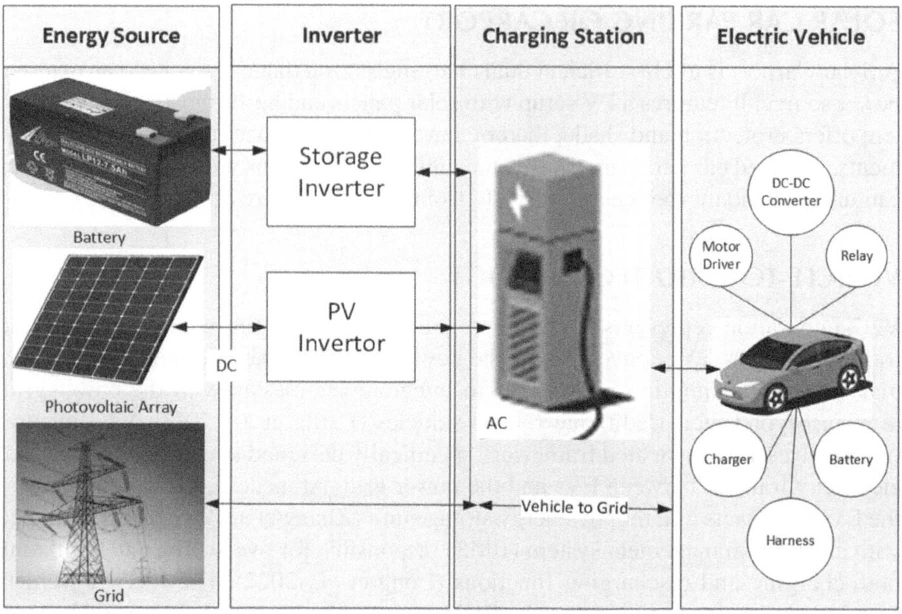

FIGURE 8.6 Vehicle-to-grid (V2G) concept.

System (DMS), continuously supervises and adjusts the energy flow to maintain grid stability (Srihari et al., 2024) (Figure 8.6).

BATTERY MANAGEMENT SYSTEM

The BMS is an electronic system designed to manage and protect rechargeable batteries, ensuring optimal performance, longevity, and safety. The BMS monitors the state of the battery, produces secondary data, and generates essential information reports. Key metrics such as state of charge, state of health, and remaining capacity are among the important parameters that the BMS tracks and calculates. Moreover, it keeps an eye on vital factors like current, voltage, and temperature. In addition to providing protection, the BMS also regulates the battery's environment by managing heating or cooling systems to maintain its optimal operating temperature range. Another essential function of the BMS is cell balancing, which ensures that each cell in a battery pack charges and discharges evenly, improving the battery's overall performance and lifespan. This system's comprehensive monitoring, reporting, and protection features are crucial for maintaining the reliability and safety of modern rechargeable batteries.

Lithium-ion (Li-ion) and nickel metal hydride (NiMH) batteries are two of the most common types of electric batteries in use today. Lithium-ion batteries utilize lithium and cobalt as electrode materials, while nickel-based batteries, specifically NiMH, use nickel as their primary component. The global shift from gasoline-powered vehicles to EVs requires a considerable increase in the use of these metals. Fraser et al. (2021) forecast global nickel demand for EVs to grow from 92 kilotonnes in 2020 to 2.6 million tonnes by 2040. Research conducted by Karabelli et al. (2020) suggests that the demand for e-mobility will boost battery production to around 1,725 GWh by 2030, with nickel serving as the primary raw material for lithium-ion batteries. Currently, nickel consumption in these batteries represents 4% of the total annual output globally.

INDIAN SCENARIOS OF EVS AND CHARGING STATION

India has set a goal to boost the share of EV sales to 30% for private cars, 70% for commercial vehicles, 40% for buses, and 80% for two-wheelers and three-wheelers by 2030. This represents an ambitious target of 80 million EVs operating on Indian roads by that year. Moreover, India is committed to ensuring that all EVs are produced domestically under the 'Make in India' initiative. In January 2025, EV sales in India experienced a remarkable rise, with a month-over-month increase of 19.4% and a year-over-year growth of 17.1%, totalling 169,931 units sold. The year 2023 witnessed a substantial surge in EV sales, climbing by 49.25% to reach 1.52 million units. While the sector remains in its nascent stages, it is progressively gaining momentum and will launch nearly a dozen new EVs in 2025, focusing on premium models, as India's EV sales rose 20% despite slowing global demand, with a 30% target by 2030.

The global transition to EVs presents new opportunities for suppliers in the automotive sector. The market for EV batteries in India is expected to rise from US$ 16.77 billion in 2023 to an impressive US$ 27.70 billion by 2028. Regarding infrastructure, as of February 2024, there are 12,146 public EV charging stations in operation across the country, with Maharashtra leading in the number of charging stations, followed by Delhi and other states (Invest India/IBEF, 2024). A recent report from the Confederation of Indian Industry (CII) (2023) highlighted the urgent need to set up at least 1.32 million charging stations in India by 2030 to support the swift expansion of EVs, necessitating over 400,000 installations each year.

OPPORTUNITIES AND CHALLENGES

Fuel savings and effect on economy

Anand (2023) forecasts that by 2030, there will be 16 million EVs on the roads, which could lead to a notable reduction in fuel consumption assuming an equivalent number of ICEVs are replaced. It is projected that the annual fuel savings from replacing 16 million ICEVs with EVs will amount to 2,196.5 million litres. The demand for crude oil is expected to decline significantly, as petrol and diesel are both produced from it. The 2,196 million litres of fuel saved by 2030 translates to approximately 30.6 million barrels of crude oil. Considering that India's oil imports are around 4.7 million barrels per day, the total savings from substituting ICEVs with 30.6 million barrels has a minimal effect on India's crude oil import expenses.

Government estimates indicate that there will be roughly 350.7 million registered vehicles on the roads in India by 2030 (CEA, 2022), which suggests that only about 4.5% of vehicles will be electric. This also shows that India is significantly behind its goal of achieving a 30% share of EVs by 2030. Given this small percentage of vehicle registrations, the potential impact on fuel savings and crude oil imports remains limited, based on the current EV sales projections.

The slowdown in economic activity associated with fuel production and usage also results in a reduction of total greenhouse gas (GHG) emissions by 0.27%. This reduction is linked to decreased petrol and diesel production due to the replacement of ICEVs. As previously noted, with EVs constituting just 4.5% of all registered vehicles by 2030, the environmental savings in terms of emissions will be minimal, and the road transport sector will keep its strong dependence on fossil fuel sources.

EVs ASSOCIATED ELECTRICITY DEMAND AND GREENHOUSE GAS EMISSIONS

Environmental burden and human health cost of EV battery charging for regulating vehicular tail-pipe emissions, India has adopted the Bharat Stage (BS) emission standards, which have been modelled on the EU norms since the year 2000 (IEA, 2023). In 2020, India adopted the BS VI standards which are similar to the Euro VI norms and have been followed by all vehicle manufacturers in the country. The GHG and non-GHG emissions from coal-fired thermal power plants and ICEVs are proven to be more damaging to human health than the emissions from other pollution sources (Thurston et al., 2021). McCubbin and Delucchi (1999) had modelled the cost per kilometre of vehicle travelled for different pollutants. While the latest studies have also estimated the externality cost of emissions, the novelty of this study is the cost burden being presented for each pollution as highlighted in the BS VI standard on motor vehicle emissions highlighted in Table 8.1, based on the severity of the impact on human health. To this end, using the global inflation rates to adjust the cost at 2023 price levels, the per-unit costs are presented in Table 8.1.

Given the per unit health costs, the total cost saved by substituting the equal amount of ICEVs with EVs and decreasing on-road air emissions as presented in Table 8.1 is estimated. On the other hand, incremental emissions are also incurred from EV charging from coal-fired TPPs. To this end, the study adopts the mortality costs provided by Chakravarty and Somanathan (2021) for the GHG and local air emissions presented in Table 8.2. The next section presents the results of the study.

TABLE 8.1
Externality costs from motor vehicles (USD/km)

S. no.	Emissions	Petrol	Diesel
1	CO	3.48	0.41
2	HC	0.50	2.24
3	NOX	2.07	6.47
4	PM	71.4	537.85

Source: McCubbin and Delucchi (1999) and Prabhu and Mukhopadhyay (2023).

TABLE 8.2
Externality cost from coal power plants for EV charging (USD/Kwh)

S. no.	Emissions	Externality cost
1	Environmental cost of mining	0.0029
2	Mortality cost of air pollution	0.0203

Source: Chakravarty and Somanathan (2021).

TECHNICAL LIMITATIONS

One of the main challenges of using solar PV in transportation is the technical limitations of the technology itself. Solar PV panels have low energy density, meaning they produce a small amount of power per unit area. This limits the amount of solar energy that can be captured and stored on a vehicle, especially when compared to conventional fuels. Moreover, solar PV performance depends on several factors, such as weather conditions, orientation, shading, and degradation. These factors can reduce the efficiency and reliability of solar PV systems in transportation applications.

COST AND FEASIBILITY

Another obstacle to implementing solar PV in transportation is the cost and feasibility of the systems. Solar PV panels are expensive to manufacture, install, and maintain, and they require additional components, such as batteries, inverters, and controllers, to operate effectively. The cost of solar PV systems can outweigh the benefits of fuel savings and emission reductions, especially for large-scale and long-distance transportation modes. Furthermore, the feasibility of solar PV systems depends on the availability and accessibility of land, infrastructure, and grid connections. These factors can vary widely across different regions and countries, affecting the viability of solar PV projects.

POLICIES AND REGULATIONS

A third challenge of using solar PV in transportation is the policy and regulation environment. Solar PV systems are subject to various laws and standards that regulate their safety, performance, and integration. For example, solar PV panels must comply with fire, electrical, and structural codes, as well as vehicle weight and size restrictions. Additionally, solar PV systems must be compatible with the existing grid and market structures, which may not be designed to accommodate variable and distributed generation sources. The lack of clear and consistent policies and regulations can hinder the adoption and innovation of solar PV systems in transportation.

SOCIAL AND ENVIRONMENTAL IMPACTS

A fourth obstacle to implementing solar PV in transportation is the social and environmental impacts of the technology. Solar PV systems have positive effects on reducing GHG emissions and air pollution, but they also have negative effects on land use, water consumption, and waste generation. For instance, solar PV panels require large amounts of land to produce sufficient power, which can compete with other land uses, such as agriculture, biodiversity, and recreation. Moreover, solar PV panels consume water for cleaning and cooling, and generate waste materials during their production and disposal. These impacts can affect the sustainability and acceptability of solar PV systems in transportation.

REFERENCES

Anand, S. 2023. India's EV revolution: Dashboard predicts 45.5% CAGR, 1.6 crore EVs by 2030. https://energy.economictimes.indiatimes.com/news/power/indias-ev-revolution-dashboard-predicts-45-5-cagr-1-6-crore-evs-by-2030/104471821

CEA. 2022. Report on twentieth electric power survey of India (Volume-I). New Delhi: Central Electricity Authority, Ministry of Power, Government of India. Retrieved from https://cea.nic.in/wpcontent/uploads/pslf/2022/11/20thEPSReportFinal16.11.2022.pdf.

Chakravarty, S. and Somanathan, E. 2021. There is no economic case for new coal plants in India. *World Development Perspectives*, 24. https://doi.org/10.1016/j. wdp.2021.100373.

Confederation of Indian Industry (CII). 2023. Roadmap for future mobility 2030: Charging infrastructure. CII-Edelman Advisory report.

Demirbas, A. 2009. Biofuels securing the planet's future energy needs. *Energy Conversion and Management* 50(9), pp. 2239–2249.

Fouquet, R. 2014. Long-run demand for energy services: Income and price elasticities over two hundred years. *Review of Environmental Economics and Policy* 8(2), pp. 186–207.

Fouquet, R. and Pearson, P.J. 2012. Past and prospective energy transitions: Insights from history. *Energy Policy* 50, pp. 1–7.

Fraser, J., Glencore Technology, and Institute of Materials. 2021. Nickel and the energy transition: The growing role of nickel in electric vehicle batteries. https://www.glencore.com

Gales, B., Kander, A., Malanima, P. and Rubio, M. 2007. North versus South: Energy transition and energy intensity in Europe over 200 years. *European Review of Economic History* 11(2), pp. 219–253.

IEA. 2021. World energy outlook 2021. https://www.iea.org/reports/world-energy-outlook-2021

IEA. 2023. Global EV outlook. Confirm. 2023. Catching up with climate ambitions. Paris: International Energy Agency https://iea.blob.core.windows.net/assets/dacf14d2-eabc-498a-8263-9f97fd5dc327/GEVO2023.pdf.

IEA. 2024a. Global EV outlook 2024: Outlook for battery and energy demand. International Energy Agency. https://www.iea.org/reports/global-ev-outlook-2024/outlook-for-battery-and-energy-demand

IEA. 2024b. Global EV outlook 2024: Outlook for energy demand. International Energy Agency. https://www.iea.org/reports/global-ev-outlook-2024/outlook-for-energy-demand

Invest India/IBEF. 2024, Feb 13. Charging ahead: The evolution of electric vehicle infrastructure in India. https://ibef.org/blogs/charging-ahead-the-evolution-of-electric-vehicle-infrastructure-in-india

Jaramillo, P., Kahn Ribeiro, S., Newman, P., Dhar, S., Diemuodeke, O.E., Kajino, T., Lee, D.S., Nugroho, S.B., Ou, X., Hammer Strømman, A. and Whitehead, J. 2022. Transport. In Climate Change 2022: Mitigation of Climate Change. https://report.ipcc.ch/ar6wg3/pdf/IPCC_AR6_WGIII_FinalDraft_Chapter10.pdf

Karabelli, D., Johansson, M. and Koskinen, T. 2020. Sustainable battery materials and resource needs for future e-mobility. *Journal of Cleaner Production* 275, p. 123054. https://doi.org/10.1016/j.jclepro.2020.123054

Kramer, B., Chakraborty, S. and Kroposki, B. 2008, November. A review of plug-in vehicles and vehicle-to-grid capability. In *2008 34th Annual Conference of IEEE Industrial Electronics* (pp. 2278–2283). Orlando, FL, USA: IEEE. doi: 10.1109/IECON.2008.4758312.

Long, W., Xiao, Z., Wang, D., Jiang, H., Chen, J., Li, Y. and Alazab, M. 2022. Unified spatial-temporal neighbor attention network for dynamic traffic prediction. *IEEE Transactions on Vehicular Technology* 72(2), pp. 1515–1529.

McCubbin, D.R. and Delucchi, M.A. 1999. The health costs of motor-vehicle-related air pollution. *Journal of Transport Economics and Policy*, pp. 253–286.

Prabhu, V.S. and Mukhopadhyay, K. 2023. Macro-economic impacts of renewable energy transition in India: An input-output LCA approach. *Energy for Sustainable Development* 74, pp. 396–414.

Prabhu, V.S. and Mukhopadhyay, K. 2024. The economic and environmental consequences of the electric vehicle transition in India. *Energy for Sustainable Development* 81, p. 101459.

Scarlat, N., Dallemand, J.F., Monforti-Ferrario, F. and Nita, V. 2015. The role of biomass and bioenergy in a future bioeconomy: Policies and facts. *Environmental Development* 15, pp. 3–34.

Schroeder, A. and Traber, T. 2012. The economics of fast charging infrastructure for electric vehicles Energy Policy, 43, pp. 136–144.

Shukla, P.R., Skea, J., Slade, R., Al Khourdajie, A., van Diemen, R., McCollum, D., Pathak, M., Some, S., Vyas, P., Fradera, R. and Belkacemi, M. 2022. Climate change 2022: Mitigation of climate change. Contribution of working group III to the sixth assessment report of the Intergovernmental Panel on Climate Change, 10, p. 9781009157926.

Smil, V. 2010. *Energy myths and realities*. AEI Press.

Smil, V. 2017. *Energy and civilization: A history*. MIT Press.

Srihari, G., Krishnam Naidu, R.S.R., Falkowski-Gilski, P., Bidare Divakarachari, P. and Kiran Varma Penmatsa, R. 2024. Integration of electric vehicle into smart grid: A meta heuristic algorithm for energy management between V2G and G2V. *Frontiers in Energy Research* 12, p. 1357863.

Thurston, G., Awe, Y., Ostro, B. and Sanchez-Triana, E. 2021. Are all air pollution particles equal? How constituents and sources of fine air pollution particles (PM2. 5) affect health. https://documents1.worldbank.org/curated/en/810141630705865331/pdf/Are-All-Air-Pollution-Particles-Equal-How-Constituents-and-Sources-of-Fine-Air-Pollution-Particles-PM2-5-Affect-Health.pdf

Ustun, T.S., Ozansoy, C.R. and Zayegh, A., 2013. Implementing vehicle-to-grid (V2G) technology with IEC 61850-7-420. *IEEE Transactions on Smart Grid* 4(2), pp. 1180–1187.

Yang, H., Li, Z. and Qi, Y. 2024. Predicting traffic propagation flow in urban road network with multi-graph convolutional network. *Complex & Intelligent Systems* 10(1), pp. 23–35.

Yergin, D. 2012. *The prize: The epic quest for oil, money & power*. Simon and Schuster.

Zhang, X., Lu, Z., Yuan, X., Wang, Y. and Shen, X. 2020. L2-gain adaptive robust control for hybrid energy storage system in electric vehicles. *IEEE Transactions on Power Electronics* 36(6), pp. 7319–7332.

9 Solar-powered urban farming

CURRENT TRENDS IN URBAN FARMING

Urban farming includes urban and suburban farming, and has surfaced as a viable approach that encourages intensive farming techniques and urban wastewater to generate food sustainably. This approach presents a unique and sustainable alternative which optimizes the available urban spaces such as rooftops, vacant lots, and vertical structures to cultivate food within city boundaries. This amalgamation of agriculture within urban environments not only lessens reliance on lengthy supply chains but also promotes a clean and sustainable environment through minimal resource consumption including fresh water and pesticides, while enhancing local biodiversity (Ebissa et al., 2024). Among various models of urban farming, vertical farming represents a cutting-edge approach to agriculture, involving the cultivation of vegetables and fruits in vertically stacked layers (Zhu and Marcelis, 2023). This method permits cultivation of food throughout the year that is unaffected by weather conditions and also reduces the necessities of agrochemicals such as herbicides, fertilizers, and insecticides often used in conventional agriculture practices. The vertical farming strategy, implemented in urban indoor high-rise buildings with climate control, aspires to significantly boost productivity and diminish environmental impact. These facilities are believed to provide various potential advantages as a clean and sustainable food source, alongside bio-security, immunity from pests and droughts, the conservation of up to 90% of water, and lowered reliance on transportation and fossil fuels (Benke and Tomkins, 2017). The hydroponics and aquaponics do not need the soil for nutrients supply but the major difference is that hydroponics only involves growing the plants in a nutrient-mixed water medium while the aquaponic system adopts an integrated approach for growing the fishes and plants together where fish waste fertilizes the plants, and the plants helps in cleaning the water for the fishes. Aeroponics is also soil-less farming but it requires nutrient rich mist as plants are grown like their roots suspended in the air to get the nutrients from nutrient rich mist medium. Unlike traditional agriculture that typically relies on single surfaces, vertical farming is growing the plants in vertically arranged levels. Nowadays commercial urban farming involves integration of hydroponics, aquaponics and aeroponics with vertical farming in formally structured modular towers to optimize yields in limited spaces.

VARIOUS MODELS FOR URBAN FARMING

Urban farming typically falls into four primary models: community farming, indoor farming, rooftop farming, and vertical farming (Despommier, 2013; Drottberger

DOI: 10.1201/9781003652229-9

et al., 2023; Jansma et al., 2024; Parameswari et al., 2024). The various methods or models are utilized depending on the available space, materials or tools, and variety of plants or vegetables produced. The most common methods employed for urban farming are hydroponics, aquaponics and aeroponics.

COMMUNITY FARMING

A community farm is a shared agricultural approach typically located in urban settings, where crops are cultivated predominantly outdoors to supply foods locally to neighbourhoods (Poulsen et al., 2017). Urban farmers commonly grow fruits plants and vegetables including long beans, okra, brinjal, pumpkins, cucumbers and leafy vegetables like spinach in community farms. This urban agricultural method highlights the collaborative efforts of urban farmers, focusing on local farming and minimizing the requirement of food through transportation (Salim et al., 2022).

INDOOR FARMING

An indoor farm is a sophisticated and complex approach where environmental factors like light, humidity, and temperature are regulated to grow plants; this controllable strategy makes indoor farming a recognized and effective urban farming model (Wong et al., 2020). Indoor farming needs a small space in vertical levels and control conditions for regular food production as compared to community farming, which often require more space and food production affected by weather conditions. It is also soil-less farming where hydroponics or aeroponics is utilized in vertical systems; most suitable for high demand like some leafy greens such as broccoli, cabbage, lettuce and high-value herbs to cater the specific needs and high market demands Ampim et al., 2022). However indoor requires high initial investment and often integrates advanced technologies like LED lighting, climate control systems, sensors, and automation (Van Delden et al., 2021), which have demonstrated effectiveness in enhancing resource efficiency, improving plant growth and crop quality, and mitigating pest-related issues.

ROOFTOP FARMING

A rooftop farm is an agricultural model that cultivates plants on the tops of buildings, utilizing neglected or underused spaces in urban environments (Sabeh, 2016). This farming approach contributes to local food production on a local level which not only reduces the dependency on conventional agriculture but also serves a practical response to achieve food security. Rooftop farming facilitates the application of farming principles and practices into densely populated urban settings where so much of roof space is available and underutilized as traditional agriculture land becomes scarce or utilized for other land use like residential and commercial purposes (Appolloni et al., 2021). Rooftop farms can be installed in small scale like individual homes, apartments to large scale like industrial or institutional buildings for commercial purposes.

VERTICAL FARMING

Vertical farming is a modified form of indoor farming where plants are cultivated in vertically arranged spaces, warehouses, and skyscrapers under a controlled indoor environment. It also employs agricultural techniques like hydroponics, aquaponics, and aeroponics, for optimizing resources use and enhancing crop production (Katoch et al., 2024). These farms can be established as small enterprises located in building halls or shipping containers or warehouses to large-scale commercial establishments with multiple vertical levels (Birkby, 2016). Vertical farms provide many advantages over rooftop farms as they utilize vertical spaces over unoccupied rooftops urban settings; this leads to relatively higher food production per square foot area. This type of farming approach also needs high initial cost through incorporating advanced technologies such as LED lighting, climate control systems, sensors, and automation (Bhuyan et al., 2023) which enables continuous crop production irrespective of external weather conditions. These methods present numerous potential benefits, including reduced water usage, shorter growing times, and decreased dependency on pesticides.

SOLAR-POWERED URBAN FARMING

Urban farming introduces an innovative model for urban agriculture which incorporates farming approaches and systems within building structures by applying various farming models such as hydroponics or aeroponics. Various studies have shown that urban spaces-integrated farming systems can cultivate leafy vegetables while using only 95% less water than traditional farming techniques. These systems provide considerable advantages in urban food security while enhancing building sustainability by improving thermal performance and integrating biophilic design. The adoption of vertical farming systems in buildings poses particular technical challenges, specifically in terms of irrigation integration and access for maintenance. Critical issues highlighted include intricate plumbing demands, the necessity for precise environmental control systems, and challenges in managing plant health in vertical setups. These challenges are exacerbated by the requirement to maintain consistent growing conditions while preserving the integrity of the building envelope and allowing for routine maintenance access (Figure 9.1).

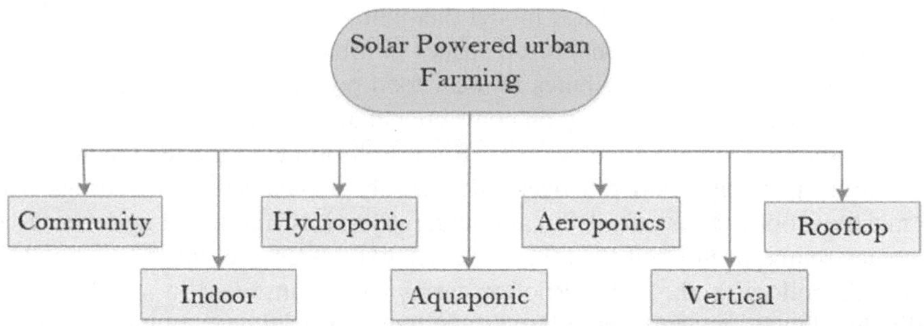

FIGURE 9.1 Solar-powered urban farming.

HYDROPONICS

Hydroponics refers to a vertical farming technique where plants are cultivated in soil-less environment, mainly water enriched with nutrients essential for the plant growth, mainly. It minimizes the risk of soil-borne diseases, mainly fungal pathogens that enhance the plant growth. Instead of relying on soil for root support and nourishment, plants are typically provided with nutrient-dense water, which is crucial for their growth, using inert mediums like cocopeat. Hydroponic systems consume 60–70% less water compared to traditional agriculture (Royston and Pavithra, 2018).

AEROPONICS

Aeroponics is another soil-less vertical farming method that involves spraying a mist containing essential nutrients directly onto plant roots. This technique requires 95% less water than conventional agricultural practices and even less than hydroponics. Thus, it serves as a viable solution for urban areas experiencing water shortages (Al-Kodmany, 2018).

AQUAPONICS

Aquaponics is an efficient approach to simultaneously cultivating plants and fish in a water-based system that closely resembles hydroponics (Fernández-Cabanás et al., 2020). This method introduces a new paradigm for the traditional aquaculture, allowing plants to grow in shallow water without soil. A key advantage of this system is the integration of plant cultivation and fish production, which facilitates nutrient exchange through the shared water between the two systems.

SOLAR PV TECHNOLOGIES

The electricity from integrated vertical farms—solar PVs or agrivoltaics is clean, renewable and less carbon-intensive making it a more sustainable option when compared to conventional electricity from grids; and any surplus energy produced during the process can be stored for future use, further enhancing the efficiency of the vertical farming. Solar-integrated vertical farming incorporates solar power directly into a vertical farming system to facilitate electricity production for various equipments like sensors and control systems and pumping for irrigation. For the efficient solar electricity production, three types of agrivoltaics technologies are available: (a) Interspersed PV arrays that involves placing solar panels in the spaces between the plant rows; (b) Greenhouse-mounted PV arrays involve placing PV modules above the crops by substituting part of the greenhouse covering with panels; (c) Stilt-mounted PV arrays, involves placing PV modules above the crops by mounting them on an open structure. Further PVs and their arrangement on the mounting structure could either be fixed to the structure (static panels) or adjustable (dynamic panels), allowing their angle to be modified based on sunlight and farming management.

BUILDING-INTEGRATED PHOTOVOLTAICS

Agrivoltaics building envelopes represent a novel concept merging agrivoltaics—using space effectively for both solar energy and agriculture—with building integration technologies. The core idea of agrivoltaics, as defined by recent studies, investigates the interplay between solar panels and plant growth. Recent research has explored diverse strategies for integrating building-integrated photovoltaics (BIPV) with farming practices in building facades, such as alternating panel layouts and semi-transparent PV integration alongside plant-growth areas. These systems can be categorized into two primary types: elevated solar panels above crops and inter-row arrangements, each providing unique benefits for energy production and crop cultivation. Research has shown that well-designed agrivoltaics systems can achieve substantial solar energy output with conventional BIPV, while also ensuring favorable conditions for plant growth.

SOLAR DRYERS

Solar dryers can be categorized according to their heating mechanism into direct, indirect, mixed, and hybrid types, as well as by airflow into natural convection (passive) and forced convection (active) systems (El-Sebaii and Shalaby, 2012). Direct solar dryers allow the product to be exposed directly to sunlight, with airflow driven by natural convection; thus, the drying rates, efficiency, and quality of the product are relatively low. These are suitable for small-scale agriculture. Indirect solar dryers are more frequently utilized to shield the product from direct sunlight. In this approach, air is heated using solar collectors and then circulated into the chamber where the product is kept. This design results in quicker drying times and improved drying quality, making it ideal for drying large quantities of food. Indirect solar drying can occur using either natural or forced convection methods. In natural convection solar dryers, the flow of heated air is generated by the thermal gradient, whereas in forced convection dryers, also known as active dryers, fans are used to move the air. Consequently, airflow rates are greater, and the drying conditions can be better regulated (Fudholi et al., 2014). Among various indirect systems, greenhouse dryers have attracted attention due to their straightforward design and ease of construction (Jain and Tiwari, 2004). Lastly, hybrid systems integrate renewable and conventional energy sources to maintain a consistent air temperature, resulting in higher-quality finished products and improved control over the drying conditions.

SOLAR COLD STORAGE

The solar cold storage system effectively utilizes solar energy for cooling perishable goods, allowing for longer preservation. The cold storage unit operates off-grid, relying primarily on power produced from Solar Photovoltaic sources, but can switch to grid power on overcast days (Karthikeyan et al., 2022). During daylight hours, the electricity generated by solar photovoltaic panels powers a vapour compression refrigeration cycle to cool a storage room and also charges a thermal energy storage

(TES) system. In TES systems, cold energy is kept in phase change materials, like water or a eutectic mixture of water and salt, and is distributed to the cold storage unit based on demand. When solar energy is unavailable, the stored cooling energy from the TES system satisfies the cooling requirements of the cold storage unit. The combination of solar photovoltaic systems and TES reduces reliance on the grid and eliminates the need for diesel generators as backup power sources.

OPPORTUNITIES AND CHALLENGES

Cost reduction

Agriculture is a sector that requires substantial energy, mainly for processing, irrigation, and mechanization. Traditionally, farmers have depended on costly energy sources such as grid electricity or diesel-powered pumps. Solar energy, however, presents a potentially cheaper alternative. By implementing solar-powered systems, farmers can significantly cut their energy expenses over time. These savings can then be reinvested into other areas of the business, like enhanced agricultural equipment, better seeds, or improved irrigation systems. Solar energy is both efficient and cost-effective. For example, farmers can use solar-powered irrigation systems to automate their watering schedules, ensuring that their crops receive the right amount of water at the right times. This leads to improved crop yield, reduced water wastage, and enhanced overall operational efficiency on the farm. Moreover, solar energy can power tractors, pumps, and processing equipment, boosting farm productivity while decreasing dependence on fossil fuels and manual labour.

ENVIRONMENTAL SUSTAINABILITY THROUGH BETTER USE OF RESOURCES

Solar energy serves as a clean, sustainable power source that has minimal environmental impact. With a rising awareness of climate change and its consequences on agriculture, many farmers are turning to solar energy to reduce their carbon footprint. For instance, solar-powered irrigation systems eliminate the need for diesel-powered pumps that emit harmful greenhouse gases. By adopting solar energy, farmers can enhance the environmental sustainability of their operations and contribute to the battle against climate change. Farmers in remote or off-grid areas frequently encounter challenges related to unreliable or costly electricity access. By harnessing solar energy to generate their own electricity, these farmers can lessen their reliance on the grid and fuel-driven generators. Solar energy is a reliable and sustainable choice for agricultural operations, ensuring that they are not disrupted by power outages or fluctuations in energy costs. Water scarcity is a significant issue for farmers, especially in regions with low rainfall. Solar-powered water pumps present a cost-effective and eco-friendly option for irrigation and water extraction. By employing these systems, farmers can irrigate their crops without relying on expensive and inefficient fossil fuel-powered pumps. Additionally, solar-powered irrigation can offer more precision, reducing water waste and ensuring that crops receive optimal hydration, even in water-scarce areas.

SUPPORT FOR OFF-GRID FARMING AND LONG-TERM BENEFITS

Electricity access is often limited or nonexistent in many parts of the world, particularly in rural and developing areas. By allowing farmers to power their operations independently from the grid, solar energy is transforming off-grid farming. Solar panels can energize everything from refrigeration units for keeping fresh produce to irrigation systems. This innovation not only helps farmers remain competitive but also creates new agricultural growth opportunities in remote regions. While the initial investment for solar energy systems can be substantial, the long-term financial gains far outweigh these costs. With reduced energy expenses and enhanced agricultural productivity, farmers can ultimately achieve a return on investment when the appropriate solar infrastructure is in place. Additionally, many countries offer subsidies or other incentives that facilitate the transition to solar energy.

ACCESS TO LAND AND IRRIGATION

Securing land for urban agriculture is a significant challenge, especially in densely populated areas. These challenges can be mitigated by partnering with local authorities to find and lease vacant lots for urban farming, creating community-owned land, and promoting zoning regulations that support agricultural practices. Furthermore, effective water management is vital for urban agriculture, particularly in regions facing water shortages. Utilizing techniques such as drip irrigation, rainwater harvesting, greywater recycling, and innovations like Smart Irrigation Systems, Cloud-based Monitoring and Control Systems, and desalination methods for using seawater or brackish water can enhance water efficiency and minimize waste in urban farming.

PEST AND DISEASE MANAGEMENT

Urban agriculture faces numerous challenges, particularly in managing pests and diseases. This problem is exacerbated by the close proximity of crops to one another and to residential properties, leading to significant yield losses. However, the sustainability of urban farming can be improved through interdisciplinary research collaborations and proactive involvement with the industry. In outdoor urban growing methods such as aquaponics and hydroponics, pest issues are similar to those found in conventional agriculture. Waterborne diseases, notably, can disseminate rapidly, potentially destroying entire crops. Research has indicated the presence of microbial pathogens in aquaponic (Dinev et al., 2023; Teoh et al., 2023) and hydroponic systems (Osdaghi, 2023).

FINANCIAL CONSTRAINTS

Financial challenges are significant obstacles to the effectiveness and expansion of urban farming initiatives. The setup and maintenance of urban agricultural systems necessitate considerable upfront investment in infrastructure, equipment, and vital supplies such as irrigation systems. Additionally, ongoing expenses, including land leasing or purchasing costs, utility bills, and labour wages, can impose extra strain on limited financial resources. This encompasses the construction of infrastructure, installation of lighting systems, establishment of operational control and monitoring systems, and the acquisition of essential equipment. For individuals aiming to enter urban farming, obtaining funding for initial startup costs and continuous operations can be especially challenging, particularly in low-income areas where access to credit and investment prospects is limited.

KNOWLEDGE AND SKILL GAPS

A notable obstacle for urban farming projects is the prevalence of knowledge and skill deficits among farmers, residents, and stakeholders engaged in agricultural activities. The complexity of these systems amplifies these challenges, making it critical to develop standard protocols and offer training in system upkeep. Therefore, to run and maintain urban farming systems effectively, it is essential to possess both basic and advanced knowledge to ensure reliable food production outcomes. Consequently, training and capacity-building programs are vital to equip individuals with the skills and information necessary for effective involvement in urban farming.

REFERENCES

Al-Kodmany, K. 2018. The vertical farm: A review of developments and implications for the vertical city. *Buildings* 8(2), p. 24.

Ampim, P.A., Obeng, E. and Olvera-Gonzalez, E. 2022. Indoor vegetable production: An alternative approach to increasing cultivation. *Plants* 11(21), p. 2843.

Appolloni, E., Orsini, F., Specht, K., Thomaier, S., Sanyé-Mengual, E., Pennisi, G. and Gianquinto, G. 2021. The global rise of urban rooftop agriculture: A review of worldwide cases. *Journal of Cleaner Production* 296, p. 126556.

Benke, K. and Tomkins, B. 2017. Future food-production systems: Vertical farming and controlled-environment agriculture. *Sustainability: Science, Practice and Policy* 13(1), pp. 13–26.

Bhuyan, S., Laxman, T., Saikanth, D. and Badekhan, A. 2023. *Advanced farming technologies for pollution reduction and increased crop productivity* (pp. 194–214). Advanced Farming Technology; Scripown Publications.

Birkby, J. 2016. Vertical farming. *ATTRA Sustainable Agriculture* 2(1), pp. 1–12.

Despommier, D. 2013. Farming up the city: The rise of urban vertical farms. *Trends in Biotechnology* 31(7), pp. 388–389.

Dinev, T., Velichkova, K., Stoyanova, A. and Sirakov, I. 2023. Microbial pathogens in aquaponics potentially hazardous for human health. *Microorganisms* 11(12), p. 2824.

Drottberger, A., Zhang, Y., Yong, J.W.H. and Dubois, M.C. 2023. Urban farming with rooftop greenhouses: A systematic literature review. *Renewable and Sustainable Energy Reviews* 188, p. 113884.

Ebissa, G., Yeshitela, K., Desta, H. and Fetene, A. 2024. Urban agriculture and environmental sustainability. *Environment, Development and Sustainability* 26(6), pp. 14583–14599.

El-Sebaii, A.A. and Shalaby, S.M. 2012. Solar drying of agricultural products: A review. *Renewable and Sustainable Energy Reviews* 16(1), pp. 37–43.

Fernández-Cabanás, V.M., Pérez-Urrestarazu, L., Juárez, A., Kaufman, N.T. and Gross, J.A. 2020. Comparative analysis of horizontal and vertical decoupled aquaponic systems for basil production and effect of light supplementation by LED. *Agronomy* 10(9), p. 1414.

Fudholi, A., Sopian, K., Ruslan, M.H., Alghoul, M.A. and Sulaiman, M.Y. 2014. Review of solar dryers for agricultural and marine products. *Renewable and Sustainable Energy Reviews* 14(1), pp. 1–30. https://doi.org/10.1016/j.rser.2013.09.020

Jain, D. and Tiwari, G.N. 2004. Effect of greenhouse on crop drying under natural and forced convection I: Evaluation of convective mass transfer coefficient. *Energy Conversion and Management* 45(5), pp. 765–783.

Jansma, J.E., Veen, E.J. and Müller, D. 2024. Beyond urban farm and community garden, a new typology of urban and peri-urban agriculture in Europe. *Urban Agriculture & Regional Food Systems* 9(1), p. e20056.

Karthikeyan, K.V., Vairavan, S., Panneerselvam, S. and Anish, R. 2022. Performance study of solar photovoltaic cold storage system using phase-change materials. *Energy Storage Research & Development*, located in search summaries.

Osdaghi, E. 2023. Pectobacterium carotovorum (bacterial soft rot). *CABI Compend* 7, p. 21913.

Parameswari, P., Ragini, M., Singh, V., NM, R., Tiwari, A.K., Belagalla, N., Pandey, S.R. and Kolekar, S.N. 2024. Vertical farming: Revolutionizing sustainable agriculture in the 21st century. *Journal of Scientific Research and Reports* 30(5), pp. 917–930.

Poulsen, M.N., Neff, R.A. and Winch, P.J. 2017. The multifunctionality of urban farming: Perceived benefits for neighbourhood improvement. *Local Environment* 22(11), pp. 1411–1427.

Royston, R.M. and Pavithra, M.P. 2018. Vertical farming: A concept. *International Journal of Engineering and Techniques* 4(3), pp. 500–506.

Sabeh, N. 2016. Rooftop plant production systems in urban areas. In Kozai, T., Niu, G., Takagaki, M., (Eds.), *Plant factory. An Indoor Vertical Farming System for Efficient Quality Food Production*; (pp. 105–111). Cambridge, MA, USA: Academic Press.

Salim, M.N., Wibowo, E.W., Susilastuti, D. and Diana, T.B. 2022. Analysis of factors affecting community participation expectations on sustainability urban farming in Jakarta city. *International Journal of Science and Society* 4(3), pp. 94–105.

Teoh, S.H., Wong, G.R., Teo, W.F.A. and Mazumdar, P. 2023. First report of *Pectobacterium carotovorum* and *Pectobacterium aroidearum* causing bacterial soft rot on curly dwarf pak choy (*Brassica rapa* var. *Chinensis*) in Malaysia. *Plant Disease* 107(11), p. 3631.

Van Delden, S.H., SharathKumar, M., Butturini, M., Graamans, L.J.A., Heuvelink, E., Kacira, M., Kaiser, E., Klamer, R.S., Klerkx, L., Kootstra, G. and Loeber, A. 2021. Current status and future challenges in implementing and upscaling vertical farming systems. *Nature Food* 2(12), pp. 944–956.

Wong, C.E., Teo, Z.W.N., Shen, L. and Yu, H. 2020. Seeing the lights for leafy greens in indoor vertical farming. *Trends in Food Science & Technology* 106, pp. 48–63.

Zhu, X.G. and Marcelis, L. 2023. Vertical farming for crop production. *Modern Agriculture* 1(1), pp. 13–15.

10 Current and future policies

SOLAR POWER PRODUCTION AND FUTURE TECHNOLOGIES

Global solar PV installations achieved another milestone in 2024, reaching 597 GW—a 33% increase from 2023, and 148 GW more than the previous year (Figure 1). While the annual growth rate decelerated from the exceptional 85% spike in 2023, it remained substantial enough to solidify solar energy's leading role in the global expansion of renewable energy. In 2024, solar accounted for 81% of all new renewable capacity added globally. In terms of global power generation, solar almost doubled its contribution over the past three years, increasing its share by 1.3 percentage points just last year to reach a 7% stake in the world's electricity mix. This growth continued to enhance the penetration of renewables and pushed the additions of conventional energy sources down to a record low.

In 2024, China again led the global solar market, installing an astonishing 329 GW, which was more than six times the capacity added by the United States, and greater than the total combined of all other top ten markets. This represented 55% of global additions, further cementing China's status as the uncontested leader in solar PV deployment (Figure 2). Despite China's dominant position, nearly all other top ten markets also reported strong growth. The US achieved a record addition of 50 GW, marking a 54% increase. India saw a remarkable resurgence, surging by 145% to 30.7 GW, reclaiming third place after dropping to fifth in 2023. This strong rebound from India forced Brazil and Germany each to fall one position in the rankings, with annual installations of 18.9 GW and 17.4 GW, respectively.

Two new countries entered the global top ten in 2024. Türkiye made a significant jump to seventh place, with a five-fold increase in annual installations, reaching 8.5 GW. France joined the top ten at the tenth position, growing by 49% to achieve 4.7 GW. These installations caused Australia and the Netherlands to drop out of the top ten. Spain, Italy, and Japan completed the rankings at sixth, eighth, and ninth place, with installations of 8.7 GW, 6.8 GW, and 6.2 GW, respectively. Together, the top ten markets accounted for 81% of global solar PV installations, while the remainder of the world contributed to the remaining 19%, collectively adding 116 GW. Clearly, this market imbalance and global distribution need to be addressed by investors and policymakers moving forward—all G20 nations, along with developing countries, should experience similar solar growth.

Cumulative solar PV capacity is projected to exceed most energy analysts' predictions by 2030. If the solar market continues on its current path, global solar installations are expected to triple in the next five years, surpassing 6 TW by 2029 in the Medium Scenario. By extending this trajectory to 2030, total solar capacity is anticipated to reach 7.1 TW by the decade's end. The Global Solar Council has established

DOI: 10.1201/9781003652229-10

an aspirational goal of 8 TW of total installed capacity by 2030, which aligns with our High Scenario forecast for that year; however, the Medium Scenario falls short of this goal, indicating that decision makers must accelerate solar deployment in their respective markets (see Policy Recommendations at p. 56). Surpassing the 7 TW milestone positions solar PV as the key contributor in achieving the COP28 Dubai objective to triple renewable energy capacity to 11 TW by 2030. Based on our forecasts, PV will comprise about 65% of the renewable capacity required to meet this target, a significant increase from its 46% share of total renewable capacity in 2024 (see Figure 6).

POLICIES TOWARDS ENCOURAGING SOLAR POWER

SDG 7 focuses on solar energy as the main source of effective clean energy. Utilizing solar energy can reduce dependence on fossil fuels and other non-renewable sources that emit greenhouse gases. Here's how solar energy supports this goal:

1. Accessibility: Solar energy systems provide a decentralized power source that can be easily accessed in remote and off-grid areas.
2. Affordability: The establishment costs for solar energy systems have significantly lowered due to technological advancements and growing economies. Thanks to reliable solar panel wholesalers and manufacturers like Sova Solar, installing solar panels is now straightforward and cost-effective.
3. Energy independence: Solar energy plays a vital role in achieving energy independence by decreasing reliance on imported fossil fuels, enhancing energy security, and minimizing vulnerability to energy price fluctuations.

SDG 11: Sustainable cities and communities

The aim of SDG 11 is to develop safe, resilient, and sustainable cities and communities. By utilizing solar energy, communities can build an empowered infrastructure that not only reduces their carbon footprint but also prepares them for future challenges. Here's how:

1. Urban carbon emission reduction: With the availability of a clean and affordable energy source through solar panels, cities can cut down their reliance on fossil fuels, which are primary contributors to greenhouse gas emissions in urban settings. This shift results in a healthier environment and a significant reduction in carbon emissions.
2. Lowering energy costs for residents: Research indicates that small to medium-sized businesses can save approximately $101,259 on energy over 20 years with solar installation. Low-income communities stand to gain the most from this.

The Paris Agreement

Climate change is a global crisis that transcends national boundaries. It is a challenge that demands international collaboration and coordinated responses at all levels. To address climate change and its adverse effects, world leaders reached a significant milestone at the UN Climate Change Conference (COP21) in Paris on December 12, 2015, by adopting the historic Paris Agreement. The Agreement establishes long-term objectives to direct all countries towards:

a. Significantly lowering global greenhouse gas emissions to maintain the increase in global temperature well below 2°C over pre-industrial levels, while striving to limit the increase to 1.5°C above pre-industrial levels, acknowledging that this would greatly mitigate the risks and repercussions of climate change.
b. Regularly evaluating the collective progress towards the aims of this agreement and its long-term objectives.
c. Providing financial assistance to developing nations to combat climate change, improve resilience, and enhance their capacity to adapt to climate impacts.

The Paris Agreement operates on a five-year cycle of increasingly ambitious climate initiatives undertaken by nations. Every five years, each country is required to provide an updated national climate action plan—known as its Nationally Determined Contribution (NDC). In these NDCs, countries outline their actions to decrease greenhouse gas emissions to meet the goals of the Paris Agreement. Additionally, they specify their plans for building resilience in response to the effects of rising temperatures. In 2023, the inaugural "global stock take" on the world's efforts under the Paris Agreement concluded at COP28, resulting in a decision on how to enhance action across all areas—mitigation, adaptation, and finance—by 2030, which included a call for governments to accelerate their transition away from fossil fuels towards renewable energy sources like wind and solar power in their upcoming climate commitments.

The International Solar Alliance (ISA) is a collaborative effort between France and India, initiated during COP21 (November 2015), aimed at significantly advancing solar energy. The ISA's objective is to establish foundational rules, norms, and standards for solar energy to enable swift and widespread adoption in nations abundant in solar resources but perceived as having high risks. The ISA is implementing tangible tools, capacity-building initiatives, and innovative financing methods. One such approach is the alignment of public policies, regulations, and pricing across countries, which mitigates uncertainties, enhances the economic feasibility of solar projects, and instils confidence in investors. The ISA aspires to facilitate the installation of over 1,000 GW of additional capacity by 2030, with projected funding of $1 trillion during the same timeframe.

Furthermore, the ISA, established in 2015 by the Hon'ble Prime Minister of India and the President of France, serves as a prime illustration of how collaborative efforts can guide the global community towards a cleaner and more sustainable tomorrow. This member-driven platform focuses on strategies aimed at creating 450 GW of renewable energy by 2030. It seeks to boost the widespread adoption of solar technologies by broadening energy accessibility, assuring energy security, and facilitating the energy transition among its member nations.

INDIAN POLICIES TOWARDS SOLAR POWER

Production Linked Incentive Scheme (PLI)

The Government of India's dedication to solar energy is expressed not only through statements but also through detailed policies and programmes. The Production Linked Incentive Scheme (PLI) for the National Programme on High-Efficiency Solar PV Modules was launched to enhance manufacturing capabilities on a Giga Watt (GW) scale, with a budget of INR 24,000 crore aimed at promoting domestic production and cementing India's status as a leader in solar energy. Under the PLI (Tranche-I), an integrated capacity of 8,737 MW has been achieved, while Tranche-II has allocated 39,600 MW of domestic Solar PV module manufacturing capacity to 11 companies, resulting in an investment of INR 93,041 crore and creating over 100,000 direct and indirect jobs.

Solar Parks Scheme

The solar parks offer suitable, developed land equipped with all necessary clearances, transmission infrastructure, and access to water, road connectivity, and communication networks. This scheme accelerates the implementation of grid-connected solar power projects on a large scale for electricity production. Each solar park is expected to have a capacity of 500 MW or more, although smaller parks may also be considered in regions where acquiring contiguous land is challenging due to difficult terrain or a severe lack of non-agricultural land. The development of Solar Parks is carried out in partnership with State Governments, their associated agencies, Central Public Sector Undertakings (CPSUs), and private companies.

India's dedication to solar power is the Solar Park Scheme, which aims to develop 50 Solar Parks with capacities of 500 MW and above, targeting a total capacity of approximately 38 GW by the year 2025–26. These solar parks serve as centres for solar energy production, drawing in investments and creating a supportive environment for solar power advancement. They play a key role in achieving economies of scale, enhancing the affordability and accessibility of solar energy. So far, 11 Solar Parks with a combined capacity of 8,521 MW have been completed, and seven more with a total capacity of 3,985 MW are partially finished. Within these parks, solar projects amounting to a total capacity of 10,237 MW have been established. The Ministry of New and Renewable Energy launched this scheme on December 12, 2014, initially proposing the establishment of at least 25 Solar Parks and Ultra Mega Solar Power Projects to achieve 20,000 MW of installed solar power capacity within five years starting from 2014 to 2015. The scope of the scheme was increased from 20,000 MW to 40,000 MW on March 21, 2017, with the goal of finishing these parks by 2025–26.

PM-KUSUM initiative

The PM-KUSUM initiative, which aims to achieve an addition of 30.8 GW in solar power capacity by March 2026, is revolutionizing the agricultural sector in India by establishing decentralized solar energy facilities, replacing diesel-powered agricultural pumps with solar-powered ones, and solarizing existing grid-connected pumps. The guidelines of the scheme mandate the use of solar modules, solar cells, motor pump sets, controllers, and the balance of the system to be manufactured domestically. Known as the Pradhan Mantri Kisan Urja Suraksha evam Utthaan Mahabhiyan Yojana (PM-KUSUM), this programme was initiated in 2019 with a total funding of 344.2 billion rupees ($4.13 billion) and targets the addition of 34.8 GW of installed photovoltaic (PV) capacity by 2026. The programme consists of three components: A, B, and C.

- Component A aims for the installation of 10 GW of solar PV capacity. Any landholders, such as farmers, cooperatives, and village committees, are allowed to hire developers or discoms to construct solar power plants on agricultural land, with a maximum output capacity of 500 kW to 2 MW per plant, and the electricity produced will be sold to discoms.
- Components B and C provide subsidies for the installation of 24.8 GW of solar PV irrigation pumps on agricultural land. Specifically, Component B aims to add 1.4 million off-grid solar PV pumps.
- Component C focuses on converting 3.5 million grid-connected pumps into solar PV systems. This transformation encompasses Individual Pump Solarization (IPS), where panels are fitted to individual pumps, and Feeder Level Solarization (FLS), which involves transforming the main power supply lines into solar PV. The level of subsidy fluctuates based on the region and the applied technology.

PM SURYA GHAR: MUFT BIJLI YOJANA (PMSGMBY)

In addition, the Rooftop Solar Programme for the residential sector and the Off-grid Solar PV Applications Programme for rural areas are enhancing the accessibility of solar energy through subsidies. This support represents not merely financial aid but a strategic investment towards a future powered by solar energy. The PM Surya Ghar: Muft Bijli Yojana (PMSGMBY), noted as the world's largest domestic rooftop solar programme, reached a significant milestone with 1 million homes powered by solar energy as of March 10, 2025. Officially launched by Prime Minister Shri Narendra Modi on February 13, 2024, this groundbreaking initiative is rapidly transforming India's energy framework. With 4.73 million applications submitted, the programme has already distributed ₹4,770 crore in subsidies to 613,000 beneficiaries, making solar energy more widely available. The programme's user-friendly financing options, which include collateral-free loans of up to ₹200,000 at a 6.75% subsidized interest rate through 12 Public Sector Banks, have also contributed to its swift uptake. Up to now, 310,000 loan applications have been submitted, with 158,000 approved and 128,000 disbursed, facilitating financial inclusion for many. With a smooth 15-day subsidy transfer process and zero electricity bills for numerous beneficiaries, the scheme not only powers homes but also empowers individuals. Each solar installation under the PMSGMBY offsets carbon emissions comparable to planting 100 trees, propelling India towards a more sustainable, self-sufficient, and environmentally friendly future.

INTEGRATION OF SOLAR POWER INTO URBAN PLANNING AND CHALLENGES

As the need for eco-friendly energy options grows, the use of solar power and various PV technologies in urban environments, such as building-integrated photovoltaics (BIPV), is advancing. Within urban settings, attention is directed towards land utilization, efficiency, aesthetics, flexibility, and sustainability. Ground-mounted and rooftop solar PV installations are composed of either mono or polycrystalline silicon panels, bifacial mono-crystalline panels, or thin-film technologies like Cadmium Telluride (CdTe) and Copper Indium Gallium Selenide (CIGS), in addition to third-generation options like organic solar cells, perovskite solar cells, and dye-sensitized solar cells (DSSCs). Solar PV systems can be either incorporated into buildings or attached to them. The most prevalent forms in urban settings are BIPV systems and building applied photovoltaics, which integrate PV modules into the architecture while providing efficient energy generation.

In order to promote local energy production rather than just consumption in cities, it is essential to incorporate solar energy considerations early in urban design and planning. The implementation of solar solutions in urban areas encounters various challenges that necessitate creative approaches. Space limitations are a significant concern (Kammen and Sunter, 2016), as densely populated regions often lack sufficient roof space and face competition for land from other urban necessities such as housing, services, and green areas. Mohajeri et al. (2016) examined the impact of different urban compactness indicators on solar energy potential for various PV installation technologies, including BIPV, solar thermal collectors, and direct gain passive solar systems, revealing that solar irradiance diminishes as compactness increases. Glare produced by reflective solar panels can negatively affect adjacent buildings, transportation infrastructure, and overall aesthetics (Etukudoh et al., 2024; Labib et al., 2016; Sreenath et al., 2021), while shadowing from tall buildings and trees can considerably impair the efficiency of solar panels (Wang, 2025; Yang et al., 2019).

The integration of solar systems into urban architecture introduces aesthetic and design difficulties, with public opposition often arising from worries about visual effects. Furthermore, urban power grids might be unprepared to accommodate the distributed energy contributions from solar systems, which calls for upgrades and modernization, alongside the need for energy storage solutions to balance supply and demand, particularly during nighttime (Etukudoh et al., 2024). In PV applications, only about one-fourth of the energy is converted into electricity, with a similar portion transformed into heat (Prakash et al., 2023). The related urban heat island effect, which elevates ambient air temperatures, can additionally lower solar panel efficiency (Adeh et al., 2019). In such contexts, innovative ventilation strategies and PV-thermal applications may need to be explored. Maintenance and accessibility also pose significant challenges, as urban solar installations on roofs and facades frequently lead to higher expenses and logistical issues (Shukla et al., 2018).

Regulatory and policy-related obstacles, such as inconsistent standards and protracted permitting processes, complicate implementation, while high initial costs and financing challenges hinder broader adoption (Kammen and Sunter, 2016; Shukla et al., 2018). Lifecycle and sustainability concerns are also crucial, particularly regarding the recycling and disposal of solar panels at the end of their lifespan (Ndalloka et al., 2024), as well as addressing their embodied carbon footprint. Lastly, merging solar systems with other urban infrastructures, like electric vehicle (EV) charging stations and water management systems, can be difficult due to space and infrastructure constraints (Shafiullah et al., 2022). These challenges underscore the necessity for thorough planning, innovative technologies, and supportive policies to effectively capture solar energy in urban settings. Figure 3 illustrates the obstacles faced by the integration of solar PV in urban environments.

SOLAR-POWERED SMART AND SUSTAINABLE CITIES

The idea of smart cities signifies a revolutionary method of urban development that utilizes technology and creative solutions to establish sustainable, efficient, and resilient urban settings. In terms of embracing solar energy, smart cities are crucial in hastening the shift towards renewable energy sources. Incorporating Solar Energy into City Development Sustainable urban planning is fundamental to the growth of smart cities. It entails embedding solar energy within the city's design and organization. This includes integrating solar installations into building architectures, designing green spaces, and optimizing solar panel placements on rooftops, parking lots, and public facilities. Zoning regulations and building codes can be developed to promote solar installation. For instance, cities can offer incentives or require solar-ready building practices and designate areas for solar farms or community solar projects. Sustainable urban planning also confronts challenges like shading in order to guarantee that solar panels capture the maximum amount of sunlight. Additionally, city planners must consider the wider effects of solar energy adoption, such as the necessity for energy storage, grid integration, and distributed energy resources. By infusing solar energy into the early stages of urban planning, smart cities can create a comprehensive strategy for energy generation, distribution, and consumption.

Smart infrastructure involves the deployment of energy-efficient systems throughout the city. This includes smart buildings equipped with solar panels, energy-efficient lighting, and advanced heating, ventilation, and air conditioning systems. These technologies not only lower energy usage but also enhance the city's overall sustainability. Furthermore, incorporating EV charging infrastructure is vital for promoting eco-friendly transportation. Solar-powered EV charging stations can help minimize the carbon emissions of urban transport and foster the growth of EV use. For public transport, smart cities can consider using electric buses powered by solar energy. The creation of efficient transportation networks and transit-oriented developments also aids in minimizing urban sprawl, reducing reliance on personal vehicles, and decreasing greenhouse gas emissions.

Utilizing technology for effective energy management digitalization and connectivity are vital components of smart city advancement. These technologies facilitate real-time monitoring and management of energy systems, improving overall energy efficiency and performance. The incorporation of IoT devices and data analytics platforms enables intelligent energy usage and grid administration. Smart meters and sensors provide crucial information on energy consumption patterns, allowing city officials to pinpoint opportunities for enhancement and optimization. This information is particularly valuable for comprehending peak energy demand and orchestrating solar energy generation in response. Demand response programmes, which motivate users to modify their energy consumption during peak times, can be executed more efficiently in smart cities. By harnessing digitalization and connectivity, smart cities can automatically regulate lighting, heating, and cooling systems to maximize energy usage and reduce costs. The application of advanced energy management systems also aids the seamless integration of renewable energy sources, including solar energy. These systems synchronize the energy produced from solar panels with real-time demand, ensuring a dependable and stable energy supply. In conclusion, incorporating solar energy into smart cities embodies a comprehensive strategy for sustainable urban development. Sustainable urban planning, smart infrastructure, and digitalization all contribute to the effective and efficient utilization of solar energy, fostering environmental sustainability, energy security, and economic prosperity. By adopting these principles, smart cities are not only minimizing their carbon emissions but are also paving the way for a more sustainable and resilient urban future.

Index

S

T

U

V

W